Beyond Cut-and-Paste

Engaging Students In Making Good New Ideas

by Jamie McKenzie

FNO Press
917 12th Street
Bellingham, Washington 98225
http://fnopress.com
http://fnopress.stores.yahoo.net//
Tel: 360-647-8759 Fax: 214-242-3894

This work contains some articles previously published .

Chapter One, "Beyond Cut-and-Paste" first appeared in *From Now On*, Vol 18|No 1|September 2008.

Chapter Two, "Beyond Mere Gathering" first appeared in *From Now On,* Vol 17|No 6|Summer 2008

Chapter Three, "The Brave New Citizen," first appeared in *From Now On*, Vol 17|No 3|January 2008.

Chapter Four, "Embracing Complexity," first appeared in *The Question Mark*, Vol 4|No 3|February 2008.

Chapter Five, "Managing the Poverty of Abundance," first appeared in *From Now On*, Vol 16|No 1|October 2006.

Chapter Seven, "After NCLB," first appeared in *From Now On*, Vol 17|No 5|May 2008.

Chapter Eight, "Photoshopping Reality," first appeared in *From Now On,*, Vol 16|No 2|December 2006.

Chapter Ten, "Gaining Attention in the Laptop Classroom," first appeared in *From Now On*, Vol 18|No 2|November 2008.

Chapter Twelve, "How to Challenge Assumptions," first appeared in *The Question Mark*, Vol 4|No 5|June 2008.

Chapter Thirteen, "The Evidence Gap," first appeared in *The Question Mark*, Vol 5|No 1|Octobe |2008.

Chapter Fourteen, "Bettering: A Synthesis Primer," first appeared in *The Question Mark*, Vol 4|No 6|Summer 2008.

Chapter Fifteen, "The Mystery Curriculum," first appeared in *The Question Mark*, Vol 4|No 4|April 2008.

Chapter Sixteen, "What Digital Age?" first appeared in *From Now On*, Vol 17|No 5|May 2008.

Chapter Seventeen, "A True Original," first appeared in *From Now On*, Vol 18|No 3|January 2009.

Chapter Twenty-One, "Reading Between Digital Lines," first appeared in *From Now On*, Vol 18|No 4|March 2009.

Chapter Twenty-Two, "Reading Across a Dozen Literacies," first appeared in *From Now On*, Vol 18|No 4|March 2009.

Chapter Twenty-Four, "One Liners, Bloggery and Tomfoolery," first appeared in *From Now On*, Vol 15|No 4|June 2006.

Beyond Cut-and-Paste
is dedicated to newly elected
President Barack Obama
who had the courage to challenge
the way things had been going
and the skill to set the United States
back on a course of decency,
reaffirming a respect for justice,
human rights and diplomacy.

About the author . . .

Jamie McKenzie is the editor of three online journals, *The Question Mark* (http://questioning.org). ***From Now On - The Educational Technology Journal*** (http://fno.org) and *No Child Left* (http://nochildleft.com).

A graduate of Yale with an M.A. from Columbia and Ed.D. from Rutgers, Jamie has been a middle school teacher of English and social studies, an assistant principal, an elementary principal, an assistant superintendent in Princeton (NJ), and a superintendent of two districts on the East Coast of the States. He also taught four-year-olds in Sunday school.

Jamie has published and spoken extensively on the importance of equipping students with the capacity to ask tough questions and make up their own minds.

Jamie now works with schools and teachers around the world to strengthen the questioning and thinking skills of students.

Late in his career, Jamie has also devoted himself to articulating a case against the damage he sees being done to schools and students by the ill-considered, untested, punitive measures that are part and parcel of NCLB — which he has renamed Helter-Skelter. He argues against fast food education and factory approaches to school improvement, suggesting instead that students be shown how to think, solve problems and invent good new ideas.

A full resume listing publication credits and a detailed career history is available online at http://fno.org/JM/resume.html.

Contents

Introduction

In an Age of Bad New Ideas (Like Mortgage Derivatives) Good Ideas Matter

2008 was a year that showed the folly of unrestricted and unchecked innovation as the financial industry crashed in a dramatic and painful manner.

Many investors woke up with huge losses, and many elders found their retirement funds seriously depleted. In some cases, retirement was no longer a viable option. In other cases, the money available to support a retirement declined in an alarming fashion. Those who had invested in common stocks found they were no longer living on fixed incomes.

What had seemed like a good new idea a decade earlier — to increase the percentage of those able to buy their own homes — turned out to have some serious flaws lurking below the surface.

Many of these new home owners were not good credit risks, and the inventive mortgage strategies designed to make houses affordable to them actually put the whole system at risk.

What does the financial crisis have to do with this book?

Too many people sign mortgage agreements and other documents without reading them carefully or understanding the fine print. Schooled on a diet of simple topical research, many citizens grow up without actually knowing how to think for themselves or wrestle with the difficult questions of life. They too readily rely upon the advice of others — people like mortgage brokers, for example, who sometimes steered customers to bad rates and bad agreements because they were paid by the lenders to do so.

This book calls for a different kind of schooling — learning that equips all students with the capacity to ask tough questions and think independently.

We are suffering from a cut-and-paste culture bred by the ease with which any of us can locate and save information. Sadly, Google and its relatives give us all a false sense of security and wisdom as we can search for something as elusive as "the truth" and Google delivers an answer in less than ten seconds.

Schools and districts committed to 21st Century Skills as well as the standards of AASL and ISTE must confront this cut-and-paste culture head-on, eliminating those classroom practices that encourage and promote such lazy thinking and research, replacing them with activities that are more challenging and more worthwhile.

What are the traits of cut-and-paste thinking?

- Scooping
- Smushing
- A general lack of original thought
- Susceptibility
- Surrender

Young people who grow up simply collecting the writing and thinking of others are not capable of the kinds of thinking suggested by the New Zealand Curriculum:

Thinking is about using creative, critical, and metacognitive processes to make sense of information, experiences, and ideas. These processes can be applied to purposes such as developing understanding, making decisions, shaping actions, or constructing knowledge. Intellectual curiosity is at the heart of this competency.

Students who are competent thinkers and problem-solvers actively seek, use, and create knowledge. They reflect on their own learning, draw on personal knowledge and intuitions, ask questions, and challenge the basis of assumptions and perceptions.

http://nzcurriculum.tki.org.nz/the_new_zealand_curriculum/key_competencies

This book tackles the challenge posed by the New Zealand Curriculum. How can schools and teachers best nurture this kind of thinking so young ones emerge from school capable of invention as well as understanding?

Chapter 1 - Beyond Cut-and-Paste
Engaging Students in Wrestling with Questions of Import

We are suffering from a cut-and-paste culture — a trend schools and teachers must combat by emphasizing questions of import.

Information is now easy to find and easy to gather.
Truth is quite a different matter.
While information and insight are often quite different, many folks are content with simple collection — gathering, smushing and summarizing the ideas and advice of others without checking for veracity.

Even though its entries are often seriously flawed, poorly writtten and amateurish, *Wikipedia* ranked as the eighth most popular Web site in the USA in October of 2008 according to *alexa.com.*

Stephen Colbert has lampooned *Wikipedia* mercilessly in segments on *Comedy Central,* pointing out that many companies and institutions have paid to improve their entries, a practice he calls *wikilobbying.* He also coined the term *wikiality* to capture the idea that reality has been replaced by *Wikipedia*'s version of reality (not always the same thing.) http://www.colbertnation.com/the-colbert-report-videos/81454/january-29-2007/the-word---wikilobbying

In a related social phenomenon, truth in many cases has been replaced by what Colbert calls *truthiness* — "to describe things that a person claims to know intuitively or 'from the gut' without regard to evidence, logic, intellectual examination, or facts." http://www.colbertnation.com/the-colbert-report-videos/24039/october-17-2005/the-word---truthiness

Making smart decisions about investments or mortgages or a life-threatening disease can be tremendously challenging in this new information landscape.

Google, ranked as the most popular Web site in the USA in October of 2008 according to *alexa.com,* gives us all a false sense of security and wisdom as we can search for something as elusive as "the truth" and trust *Google* to deliver an answer in less than ten seconds.

Enter "the truth" in *Google,* elect "I'm Feeling Lucky" and with little hesitation or needless strain or effort we have our answer. You can find "the truth" at http://www.thetruth.com/ according to Google.

Would that it were so easy!

Eliminate topical research rituals

The first step in fighting against simple cut-and-paste thinking is to gather all teachers together to discuss and adopt a school-wide policy outlawing the assignment of topical research projects.

Students in this school will conduct research on questions of import that require that they make answers rather than find them. We will no longer assign topical research or accept papers that are little more than a rehash of other people's ideas and thinking.

This policy means an end to long scrolling lists like the one below:

Go find out about . . .

Captain James Cook
Captain Matthew Flinders
Captain George Vancouver
Captain William Bligh
Fletcher Christian
Breadfruit
The mutiny on the Bounty
Tahiti
New Zealand
Australia
Navigation
Discovery

This kind of list is an invitation to copy-and-paste, scoop and smush. The lack of defining and challenging questions to shape the research permits a HUGE range of activities. The mere gathering of information is pretty much guaranteed by going topical. The implied value of such gathering is the mistaken notion that one gains in understanding as one's piles of information grow in size. Sadly, it is possible to lose ground and find oneself fogged in by such collections - the "poverty of abundance" covered in Chapter Five of this book.

Replacing topical research with questions of import

Questions of import usually require that students wrestle with difficult challenges and build their own answers rather than relying upon the thinking of others.

Example: Which of the following captains was the best at navigation?

- Captain James Cook
- Captain Matthew Flinders
- Captain George Vancouver
- Captain William Bligh

The above question requires the collection and weighing of evidence to substantiate a well-considered judgment. Such a comparison challenges the student at the top of Bloom's *Taxonomy* on the skill of evaluation. It mirrors the type of demanding reading comprehension questions found on the NAEP Test (National Assessment of Educational Progress).

What causes the main character to do _____? Use evidence from the story in your response.
How do you think the character's actions might be different today? Support your response with evidence from the story.
Source: Chapter Two of the *Reading Framework for the 2005 National Assessment of Educational Progress* at http://www.nagb.org/pubs/r_framework_05/ch2.html

For students to build an answer to this question about the relative navigational skill of the captains, they must go past the simple judgments and opinions to be found in *Wikipedia*, so-called professional encyclopedias and even some national library sites. Reliance upon secondary sources means the student need not wrestle with evidence and judgment.

According to material written about Captain James Cook on the Canadian National Library site (but recently retired to archives):

Many believe that James Cook was the greatest ocean explorer ever to have lived, and that he mapped more of the world than any other person. It cannot be denied that he combined great qualities of seamanship, leadership and navigational skill.

3

If the student merely copies and pastes this section, it is unsubstantiated opinion, strongly worded but weakly supported.

To build their own answers, students must collect specific examples of the navigational challenges each captain faced and ultimately passed or failed, checking historical records such as ships' logs and records from the Admiralty. For example, when Bligh was abandoned in a small boat by the mutineers and managed to sail more than a thousand miles across the open Pacific, what does that prove?

For each captain, the student must consider the following questions (among others), gather the facts and then compare and contrast the records of each:

> Did he know how to use all the best instruments of his time?
> Did he keep a careful log?
> Did he usually know where they were?
> Did he ever get lost?
> Did he seem to know what he was doing?
> Did his ships have to wander around very much?
> Did he stay clear of known hazards?
> Did he know how to make the best of prevailing winds?
> Did he know when to ask for directions?

This type of research is much more demanding than copy-and-paste research, but ultimately more rewarding and more empowering because it equips students to wrestle with the real issues of their own lives.

In addition to the New Zealand Curriculum quoted in the introduction to this book, many other countries, states and provices are now stressing the importance of original thought. Three prominent organizations in the U.S. have taken a strong stand on this issue.

21st Century Skills

Schools and districts that have adopted the 21st Century Skills model for schooling will find this kind of research on target, especially for the sections on Information Literacy, Media Literacy and ICT (Information, Communications & Technology) Literacy:

> Using digital technology, communication tools and/or networks appropriately to access, manage, integrate, evaluate, and create information in order to function in a knowledge economy

Using technology as a tool to research, organize, evaluate and communicate information, and the possession of a fundamental understanding of the ethical/legal issues surrounding the access and use of information.

http://www.21stcenturyskills.org

AASL Standards for the 21st Century Learner

New standards from the American School Librarians call for students to do more than gather information when they research. They set the expectation that students develop good new ideas, not just scoop and smush, copy and paste.

Learners use skills, resources and tools to:

1. Inquire, think critically, and gain knowledge.
2. Draw conclusions, make informed decisions, apply knowledge to new situations, and create new knowledge.
3. Share knowledge and participate ethically and productively as members of our democratic society.
4. Pursue personal and aesthetic growth.

AASL Standards for the 21st-Century Learner
http://www.ala.org/ala/aasl/aaslproftools/learningstandards/
standards.cfm

NETS for Students 2007

ISTE® (the International Society for Technology in Education) has adopted standards that also stress the importance of originality:

Students demonstrate creative thinking, construct knowledge, and develop innovative products and processes using technology. Students:

A. apply existing knowledge to generate new ideas, products, or processes.
B. create original works as a means of personal or group expression.
C. use models and simulations to explore complex systems and issues.
D. identify trends and forecast possibilities.

Chapter 2 - Beyond Mere Gathering
Converting Social Networking
into Collaboration and Synergy

Proponents of Web 2.0 and social networking often make claims for the benefits of this popular (supposedly new) phenomenon that include the likelihood of collaboration and group problem-solving as students from around the world team to take on important challenges. While collaboration might result if the activities are structured in ways that produce those results, decades of school and corporate efforts suggest that quality is unlikely to result from throwing folks together in groups while leaving issues of process to happenstance. This article outlines ways to increase the productivity of groups, whether they be face-to-face or remotely connected through Web 2.0.

Web 2.0 is not so new!

Social networking is not new, actually, and might be traced back two decades to the development of the Well. Nor is online collaboration new, as Judi Harris' *Activity Structures* were introduced in 1995 and all involved social networking and collaboration.

Those who rush to embrace the latest version of social networking rarely mention these earlier efforts or the lessons we might learn from their struggles, victories and disappointments. While we have some software advancements, the value of involving students in social networking remains directly tied to the scaffolding, skill building and mentoring provided by teachers. There is nothing automatically worthwhile about exchanging messages online. This is just as true in 2009 as it was in 1995. Gossip and blather are equally senseless online or offline.

Some important definitions

Gatherings do not necessarily collaborate. For decades prior to the Internet, schools devoted a great deal of energy to strategies such as Cooperative Learning and group problem-solving, mindful of traps groups may fall into such as "group think" — a tendency of majorities to squash unusual or creative thought. This trap is captured well by the saying that "a camel is a horse created by a committee."

It takes a great deal of scaffolding for teams to learn the skills of productive problem-solving, so those who hope to mine the potential of social networking and Web 2.0 should make sure that students are well versed in these skills. The new technologies may be an improvement on the MUDs (Multi-User Dungeon, Domain or Dimension) of the past, but they mainly put folks in touch. They do not make them collaborative, productive or brilliant thinkers.

Productive collaboration requires more than the exchange of ideas and suggestions. For a team to reach the synergy level in which 3+3=13 — an inventive surge of productivity and imagination that leads to high quality and imaginative production — the players must all have some grounding in the group strategies that nurture that kind of exchange and invention.

When a jazz trio or quartet is playing with great synergy, we say they are "really cooking." They may be improvising in magical ways, but it turns out that such improvisation and magic must be grounded upon a firm foundation of musical basics.

In researching "improvisation" for *Learning to Question to Wonder to Learn*, I found an illuminating study explaining how jazz musicians learn to improvise: Paul F. Berliner's *Thinking in Jazz* (1994) University of Chicago Press. The implications for thinking and group work are as follows:

> Ironically, the thinker must usually acquire a solid foundation in the thinking of the sages, the theories of the experts and the beliefs of the academy in order to build something new and worthwhile. Invention rarely thrives on ignorance. Breakaway thinking needs a wall of ideas to push off against.

> In the world of jazz, young performers must master a repertoire of chord progressions and harmonies so that they can count on them as structures around which and through which they might weave more magical variations.

Sadly, the challenge of equipping students or any group with the process skills required to do good work is often overlooked or paid lip service. We have seen a succession of online technologies that promised great inventiveness and teaming, going back to the bulletin boards and online cultures of the 1990s, but some of these delivered results that hardly matched the promises.

Models that work

Over the years, smart schools have turned to the work of thinkers such as Peter Senge, Edward de Bono and others to acquire the methods to convert gatherings and meetings into something worthwhile and productive. Most teachers have sat through enough unproductive meetings to know that getting folks together does not automatically result in productivity.

Senge, P. *Schools That Learn: A Fifth Discipline Fieldbook for Educators, Parents, and Everyone Who Cares About Education.* New York: Doubleday, 2000.

In 1992, *FNO* published an article summarizing Senge's ideas from his book, *The Fifth Discipline.* These ideas are pertinent today for anyone hoping to move past mere online gatherings to productive synergy and group invention.

> **Team Learning** — Senge stresses the importance of teaching groups to work open-mindedly together, exploring ideas and possibilities as well as arguing about them. He differentiates between discussion on the one hand, which is generally adversarial, and dialogue on the other hand, which is collaborative.   "The Fine Art of Paradigm Busting and Flexadigm Breeding" FNO, November, 1992 http://www.fno.org/FNONov92.html

Chapter Fourteen of this book is devoted to synthesis and the skills that enable individuals and teams to invent new ideas, solutions or ways of operating. "Bettering: A Synthesis Primer" outlines a professional development program for teachers to acquire these skills usually neglected during teacher preparation so they might turn about and equip students with those skills.

De Bono's *Six Thinking Hats* outlines a process that students may employ face-to-face or online to enhance the quality of group production, as each of the six colored hats stands for a kind of thinking that a group might do while trying to come up with good new ideas.

When the strategies of Senge and deBono are combined with the dozens of procedures outlined in Michalko's *Thinker Toys: A Handbook of Creative-Thinking Techniques* (2nd edition) Ten Speed Press, 2006, the likelihood of productive collaboration is greatly increased.

Informal polls of teacher audiences during the past year revealed the problem that few educators have been given much professional development or training in synthesis or collaborative work.

The value of the moderator

Many groups perform better, face-to-face or online, when there is a skilled moderator or facilitator to stir the thinking, elevate the quality of discourse and remind the group of effective process.

Without a trained facilitator, groups can easily lapse into bull sessions or rambling exchanges that are additive rather than constructive.

Members show up at different times, add their "2 cents worth" and then sign off. We end up with a collection that may amount to very little of value.

As Andrew Keen warns in *The Cult of the Amateur: How today's Internet is killing our culture*, Web 2.0 carries with it some presumptions that include a devaluing of expertise along with an inflated sense of the wisdom of crowds and amateurs.

Online chitchat

This phenomenon persists whether we look at the 2008 version of the *Well* or online comments at the *New York Times*. In suggesting a new policy approach to Pakistan, one reader offered the following:

> July 28th, 2008 7:43 am
> cheaper solution ... we need to reduce our nuclear arsenal. we could cheaply and easily drop several dozen on Pakistan, to include the tribal areas. this is acceptable under international agreements (including the Geneva conventions) because (1) Pakistan is a proven nuclear power and (2) Pakistan is a continuing supporter of the Taliban who conducted a sneak attack the USA.
>
> Legal ... and cheap !!!

Engaging students in online chitchat does little by itself to advance their learning, just as such exchanges of opinions face-to-face in a social studies class would be of little value if the teacher did not require that students substantiate and test those opinions against standards of evidence, logic and decency. Being connected locally or internationally does not lead automatically to insight.

Taking an uncritical view of wikis is a related issue for schools to consider.

When wikis fail

Heralded as one of the wonders of the Web 2.0 world, wikis invite all of us to add our "2 cents worth" to travel sites, encyclopedias and pages devoted to special topics such as emerging technologies.

Some of these wikis prove highly valuable, as customers and clients of hotels and restaurants, for example, share their miserable experiences with touted destinations and protect us from some of the marketing hype so long employed by tourist attractions and their PR firms.

Other wikis prove to be just a flash in the pan, as initial enthusiasm wanes and a half finished collection of pages launched by a graduate class sits collecting digital dust, never updated, never maintained and rarely visited.

Wikipedia, an encyclopedia generated for the most part by anonymous contributors like you, me and the seventh grade daughter of our next door neighbor, can prove illuminating on certain topics. I first learned of a Pentagon coined term "perception management" from *Wikipedia* and found to my relief that the words quoted at *Wikipedia* did, in fact, appear on a Department of Defense document. Never having been a fan of so-called professionally developed encyclopedias, I am open to the possibility that a wiki may provide some new insights not available from those antique and well combed data sources, but there can be several serious problems with wikis.

Problem One - Quality, reliability and language

During the past year, I have modified the Wikipedia definition for "educational technology" several times because on first reading, it seemed stuffy and difficult to understand, apparently written by someone in higher ed who was quite fond of terms like "educational technologist" and highly technical terms from psychology and learning fields that might make sense only to those with advanced degrees. It was verbose, abstruse and bewildering. It was not illuminating.

With the hope of making this definition a bit more helpful to the average person or layperson, I added some sentences, deleted some phrases and changed others. Many of those changes remain today, but the resulting article is a mishmash of clashing writing styles and theories. It suffers from a lack of clarity and coherence. Take a look for yourself at the current article for educational technology and see what you think. http://en.wikipedia.org/wiki/Educational_technology

I also found problems of interpretation in an article about Matthew

Flinders, a British ship captain who was the first known European to circumnavigate Australia. Important details were left out of his life story, like his attempt to bring his wife to Australia in conflict with Admiralty regulations and his conflict with the French Governor of Mauritius that led to his long term captivity on the island. History is often unreliable when written by so-called professional historians, but wikihistories can compound these problems.

Problem Two - Bias and balance

During the past year, I also made some additions to the entry on *No Child Left Behind*, an educational law that has been extremely damaging to schools and children in the U.S.A. Alarmed by what I had witnessed, I created a Web site criticizing the law at http://nochildleft.com. The *Wikipedia* article tries to summarize arguments for and against the law but ends up with a mishmash of sections. The fate of NCLB's re-authorization slated for 2007 but delayed into 2009 by growing opposition in Congress is poorly covered. There is little mention of the argument now taking place over the law's future as covered in "Democrats Offer Plans to Revamp Schools Law" in the *New York Times*.
http://www.nytimes.com/2008/06/12/us/12education.html
This concern about bias and balance was further strengthened when I read articles documenting how powerful groups like Microsoft had been paying folks to change entries. "Lifting Corporate Fingerprints From the Editing of Wikipedia" in the *New York Times*.
http://www.nytimes.com/2007/08/19/technology/19wikipedia.html
Steven Colbert coined the term *wikilobbying* to capture this trend:

> This is the essence of wikilobbying, when money deter-
> mines Wikipedia entries, reality has become a commodity.
> Steven Colbert speaking on *Comedy Central*.

Problem Three - Currency

Finally, having tracked some issues over time in *Wikipedia*, I see that articles and entries sometimes suffer from neglect and inattention. There may be no commitment to updating and revision. In the case of the entry for *No Child Left Behind*, for example, as mentioned earlier, reauthorization of the law has stalled in Congress for more than a year, but there is no comment about that stall nor comments about which presidential candidates have either supported or opposed the law. As important news stories surface about the law or the administration's actions, failings or inaction, few of these are appearing on the list at Wikipedia because no one is taking responsibility for updating.

When the Department of Education issued a report judging the Reading First Program "Ineffective," there was no mention of this report until I added it.

In too many cases, this type of updating is left to happenstance. There seems to be no one entrusted with currency.

As this chapter was being written in July of 2008, the list of references for No Child Left Behind had 58 entries, only one of which was for the year 2008.

Wikipedianism

The *GetWiki* site uses the term "wikipedianism" to capture some of these problems:

> *Wikipedianism* on this page could include quickly outdated links and text, multiple links to pages with similar content, poorly-written passages, plagiarized and/or false content, and other Wikipedia-centric formatting, titles, and system code alien to much of the authentic WikiSphere.

MySpace, Facebook and Second Life

Back in 1997 Sherry Turkle explored a wide range of issues raised by the online experiences of teenagers in *Life on the Screen: Identity in the Age of the Internet*. The current generation of social networking software may be more attractive and more popular than those she studied, but some of the same problems and issues persist.

What is a friend?
What is a true friend?

On http://biznik.com, a social networking site for independent business types to which I belonged for a year, one member posted an article, "Why Are Social Networks Popular?" in which he asks, "Why is *Facebook* so popular? Can someone who has 2000+ "friends" on a social network really be capable of sustaining meaningful relationships with such a high volume of people?"

Circumspection

Those of us who have experimented with social networking sites have sometimes found our definitions of "friend" stretched and even strained. Friend requests from total strangers are an intriguing aspect of the experience. Sometimes these are real people. Sometimes they are frauds or robots. Sometimes they are good company. Sometimes bad.

I noted with interest that most people of my generation recommended by *Facebook* because of having attended the same schools as I had set their profiles to "private" so that pictures and profiles could not be accessed without the person accepting you as a "friend." It is a trend I have noticed over the past two years at both *MySpace* and *Facebook* even with much younger people, as circumspection has for many people replaced the early rush of excitement when strangers came calling by the dozens.

Early on, many young enthusiasts shared remarkable photographs of drunken parties and personal diaries that confessed all kinds of noteworthy exploits many would prefer to keep private. It was titillating — like digital skinny dipping — but then some prospective employers ruined the fun by browsing these profiles as part of the selection process as reported by the *New York Times* in this 2006 article, "For Some, Online Persona Undermines a Résumé."
http://www.nytimes.com/2006/06/11/us/11recruit.html

Early enthusiasm for wide open spaces shifts to circumspection and a new found appreciation of privacy.

Beware the shallow waters!

Some call for a transformation of schools and classrooms to take advantage of these new digital opportunities, but these pleas eerily echo the educational rhetoric of the 1960s, the 1970s, the 1980s and the 1990s. Ddéjà vu all over again. New century but similar agendas repackaged with new electronic threads. Each wave of new tools brings a fresh wave of calls for transformation. Schools sometimes appear to suffer from short term memory loss. *Twenty-first Century Skills* seem much like the lists touted by reformers in the 19980s.

Some of these folks are out of touch with history. They seem unaware of previous attempts to transform classrooms. Being ahistorical is dangerous. Santayana warned, "Those who cannot learn from history are doomed to repeat it." Note June 1999 *FNO* Article "Beware the Shallow Waters! The Dangers of Ignoring History and the Research on

14

Change in Schools." http://www.fno.org/jun99/teach.html

While many schools could use a good strong dose of progressive change so that students would learn to invent, collaborate and think for themselves, this challenge has been around for decades and some of us have been working on it for 40 or more years.

The effort extends well back into the first half of the previous century when John Dewey led a movement to emphasize acting, thinking and problem-solving ("My pedagogic creed"). But the effort has always been stymied, not by a lack of technologies, but by cultural and social forces as parents and communities have exerted pressure for schooling that conforms with their notions of sound learning. These forces often exert their influence through testing programs that reward a narrow definition of teaching and learning.

Previous efforts have also been hampered by a lack of attention to the nuts and bolts of collaborative production. Freestyle learning may be attractive on the surface but is rarely sustainable.

Before hopping on the social networking bandwagon and urging all students to blog, wiki and network, schools might be wise to focus rigorous attention on questioning, inventing, collaborating and powerful writing in face-to-face situations with close attention to scaffolding and skill building. Adding distance and digital to the mix may be seductive but diversionary.

Chapter 3 - The Brave New Citizen

Thomas Jefferson, among others, emphasized that the vitality of a democracy depends upon the education and participation of its citizens. Note the *NCSS Standards*:

> In essence, social studies promotes knowledge of and involvement in civic affairs. And because civic issues — such as health care, crime, and foreign policy — are multidisciplinary in nature, understanding these issues and developing resolutions to them require multidisciplinary education. These characteristics are the key defining aspects of social studies.

NCSS Standards http://www.socialstudies.org/standards

New technologies promise all kinds of miracles like stronger thinking and better writing, but it turns out that many of those promises amount to *Fool's Gold*, unless good teachers and good schools combat much of the marketing and pressure to substitute templates, wizards and short-cuts for careful research, logic and questioning.

These technologies bring a mix of peril and promise as they may promote and nurture desirable citizenship behaviors or may do the very opposite, spawning a generation content with the glib, the superficial and the well-packaged.

We must be on guard against the onset of "mentalsoftness" characterized by a preference for platitudes, near truths, slogans, jingles, catch phrases and buzzwords as well as vulnerability to propaganda, demagoguery and mass movements based on appeals to emotions, fears and prejudice.

Smart uses of new technologies such as mind-mapping software combined with strong questioning and the pursuit of "difficult truths" can serve as antidotes to the disturbing cultural drift brought on by an uncritical embrace of all things new and digital.

Whither?

A cartoon by Michael Leunig, "Vasco Pajama in the strait of a thousand lighthouses" provides a vivid image of the challenge facing schools and social studies teachers as they struggle with the impact of new technologies and new media on society and their students.
http://www.curlyflat.net/vasco.jpg
Schools must sort through the claims, predictions and threats made by self-proclaimed visionaries, evangelists and futurists who welcome the social changes with little real consideration of the impact these changes will have on the nature of social and political discourse, let alone citizenship and political behavior. They tend to exaggerate the benefits and ignore the risks. They also tend to contradict each other and point in many different directions like the light houses in Leunig's cartoon. It is easy to understand how Eric Hoffer might have said that "Guru is a word for those who cannot spell charlatan."

Leunig's character, Vasco Pajama, as he seeks truth, finds the task daunting, the beacons bewildering rather than enlightening. So will our students find the search difficult as they contend with media outlets that often distort the truth rather than reporting it. Virtual reality is a daily offering on many of these outlets.

Promises, promises, promises

For two decades now we have been hearing that computers will transform classrooms and bring amazing benefits to students in the form of improved performance, yet the evidence to support these claims is scarce.

- Better thinking
- Better writing
- A global view
- Collaborative students

Possibly because attention has focused on equipment rather than program and professional development, few of these promises have been delivered, but this article will argue that smart uses of new technologies can enhance students' thinking skills while building the foundation for strong citizenship behavior.

Every student a thinker!
Every student a citizen!

Most social studies teachers would agree that programs should seek to develop the thinking skills of all students so they are capable of making up their own minds about the pressing issues of their times. Few would settle for the notion that mere cutting and pasting of ideas will suffice. Imitation and parroting of others' beliefs makes a poor and even dangerous foundation for a democratic society as such citizens are easily swayed by the emotional appeals of manipulative leaders and usually quite impatient with careful analysis.

The wisdom of the crowd can easily shift to the violence of the mob.

What is a thinker and how could new technologies strengthen or weaken citizenship and thinking?

A thinker prefers to figure out independently what to think and believe. Consulting others is fine, but copying their thinking is a kind of surrender.

The thinker will test the claims and ideas of others to see if they stand up to scrutiny. How reliable is the evidence used to bolster those claims? How coherent is the logic used?

In an age of sound bites, mind bytes, eye candy and mind candy, the thinker is guarded and skeptical. Back in 1961, William J. Lederer published *A Nation of Sheep*. Not a pretty picture. He also coauthored *The Ugly American*. Fortunately, a thinker is unlikely to join the herd, the crowd or the mob when it comes to puzzling through the crucial issues of our times. Perhaps that is why NCSS standards emphasize action.

Traits of a good citizen?

Not everybody would agree on the traits of a good citizen. Some would focus on loyalty and obedience while others might stress dissent and activism. Some traits likely to win broad acceptance:

- Involved?
- Knowledgeable?
- Logical?
- Temperate?
- Tolerant?
- Empathic?
- Questioning?

- Open-minded?
- Brave?
- ?

What would appear on your list?

The not-so-good citizen

Even though well-meaning social studies teachers might disagree
on the list of desirable traits, the NCSS Standards do provide guidance
and some clear statements.

For sure, most would find the image below in dramatic conflict
with the kinds of thinking expected of a leader or citizen. This kind
of citizen is confident and sure of himself - so much so that he has
stopped checking the facts and has begun to operate on faith or ideology rather than grounding decisions in reality.

"The Corporate Head"
http://questioning.org/LQ/coverlq2.jpg
Terry Allen (sculptor)

Good citizens and leaders ask powerful questions.
But sometimes it takes courage as well as skill.
Questions are not always welcomed by bosses or leaders. Those
in power sometimes challenge the loyalty or the patriotism of those
who challenge the current policy. The cost of questioning can be
shame, humiliation and the loss of one's job or future. Throughout
history we have seen many examples of questioners who paid a heavy
price, whether it be Thomas More's questioning of the King's marriage, Socrates teaching the young to question or various whistle-blow-
ers who lost their jobs thanks to questioning.
The Emperor's New Clothes is a classic story illustrating the reluctance of many members of a court to challenge the Emperor's wisdom
or taste in fashion. Healthy organizations reward questioning.

Smart use of new technologies

Given the challenges outlined earlier, social studies teachers
should focus on strategies likely to equip students with the thinking
skills to resist the onslaught and create well grounded personal views
and positions. Some of the more promising possibilities are shown in
the diagram on the next page and outlined in the rest of this chapter.

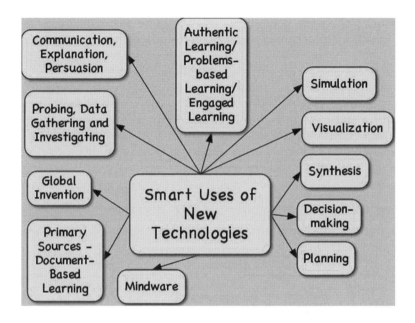

Mindware

Mind-mapping software like Inspiration™ or Smart Ideas™ serves well to map out a research strategy and then gather pertinent findings in a coherent manner.

A student researching the character of Matthew Flinders, an Australian who was the first European to successfully circumnavigate the continent in the late 1700s might produce a diagram like the one shown, with the essential question at the center and a series of suppositions surrounding it to be proven or disproven by checking primary sources like correspondence.

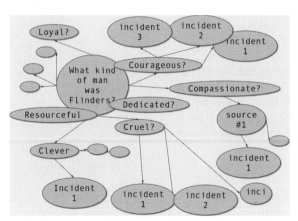

The student might also employ Inspiration™ to make a collection of images (Paintings and

Statues) depicting the man, some of which would be more accurate than others. Such a collection is not presented here because many are copyrighted.

Inspiration™ serves well when a student is picking a city in which to live or a leader. This kind of decision-making is at the top of Bloom's Taxonomy — the skill of evaluation — making a wise choice based on criteria and evidence. For each question, the student or team of students must decide what traits they want in a city or a captain. For each trait like climate (city) or navigational skill (captain) they must then pose "telling" subsidiary questions:

What is the average annual rainfall?
How hot does it get in July?
How many sunny days are there in a year?

Did he know how to use all the best instruments of his time?
Did he keep a careful log?
Did he usually know where they were?
Did he ever get lost?
Did he seem to know what he was doing?
Did his ships have to wander around very much?
Did he stay clear of known hazards?
Did he know how to make the best of prevailing winds?
Did he know how to maneuver during a sea battle?

Once posed, these questions guide the student's research and organize the findings.

For a more detailed explanation of these strategies, note the article "Inspired Investigations" at http://www.fno.org/jan03/inspiring.html

22

Authentic Learning

In many countries, social studies teachers are under increased pressure to meet challenging objectives set by state or provincial curriculum standards. To meet these thinking, problem-solving and communicating standards, it pays to involve students in seeing how these skills are practiced outside of school in a variety of organizations contending with social issues. As much as possible, it makes sense to involve them in such work, either through internships, visits, interviews or simulations. Alternatively, much of this work can also be staged as historical simulations calling for decision-making set in the past.

By employing such learning strategies within a real world context, students sharpen their abilities while gaining an appetite for the work at hand. Because they are rooted in the here and now, young ones find the challenges invigorating and intriguing. Caring about the tasks, they invest to a greater degree and emerge with a firmer and deeper grasp of the key concepts. If properly staged, historical simulations can also deliver passionate connections to provoke deep learning.

Such is the promise of Fred Newmann's concept of "authentic teaching" that involves students in "authentic intellectual work" — often outside of school. Instead of busy work — repetitive tasks that require little thought and involve mere scooping, smushing, memorizing and regurgitating Newmann's approach immerses students in challenges that demand imagination, resourcefulness, persistence and stamina.

Even though some of these activities may be staged or simulated, they still pass the test of authenticity because they meet the following criteria:

- They are rooted in issues, challenges or decisions that people face in the world.
- They are genuine.
- The act of wrestling with these challenges is purposive - saturated with meaning and significance.
- A student can see a payoff in the future for work well done and skills acquired.

These concepts are outlined in "Five Standards of Authentic Instruction" by Fred M. Newmann and Gary G.Wehlage
http://www.learner.org/channel/workshops/socialstudies/
pdf/session6/6.AuthenticInstruction.pdf

"If you were making recommendations to a local, state or federal government agency to address one of the following problems, what would be the five most important actions steps you would urge?"

acid rain	global warming	urban decay
violent crime	drunken driving	smog
traffic congestion	water pollution	declining fish harvests
endangered species	unemployment	AIDS
racial conflict	illegal immigration	gangs
voter apathy	hunger	homelessness

For more on this approach note the article "Teaching Social Studies Authentically" at http://fno.org/sept07/soc.html

Simulation and document-based learning

High school students are asked to respond to the following:

Some historians suggest that Louis XV was brought down in part by hunger and starvation. Imagine you and your team members are advising the King in the years from 1785-89. How should the King address the issues of hunger and starvation as well as taxation and wheat speculation?

In order to gain background on this challenge, they may turn to electronic sources such as *The Project Gutenberg EBook of The French Revolution*, by R. M. Johnston at http://www.gutenberg.org/files/19421/19421-8.txt.

But popular resentment, the bitter cry of the starving, applied the same name to all of them:

. . . from Louis XV to the inconspicuous monk they were all accapareurs de blé, cornerers of wheat. And their profits rose as did hunger and starvation. The computation has been put forward that in the year 1789 one-half of the population of France had known from experience the meaning of the word hunger; can it be wondered if the curse of a whole people was attached to any man of whom it might be said that he was an accapareur de blé?

As they read about taxation in France coupled at the time with speculation driving up the price of wheat, they begin to formulate changes in policies that might have saved the King from the guillotine. The king lacked foresight. Students have the advantage of 20-20 hindsight. Hunger seems especially real to them because they began with a visit to the food bank. If the King had made such visits, perhaps he would have been more alert to his danger and responsibility.

They can then create persuasive multimedia products to help show the King what his choices are.

Other Strategies

In the original version of this chapter, when it was presented as a speech to the NCSS, there were examples presented for each of the following strategies:

- Communication and Persuasion
- Probing, Data-Gathering & Investigating
- Visualization & Synthesis
- Planning and Scenario-Building
- Computer Modeling
- Global Invention

These can be found online at http://www.fno.org/jan08/new.html

Chapter 4 - Embracing Complexity

As outlined in the previous chapter, we live in times awash in simplicity and simple-minded thinking.

But life is not simple. Nor are the challenges and issues facing us all, yet our culture seems to thirst for simple answers to complex problems.

Schools must engage students in research and learning requiring them to construct answers and make up their own minds. Teachers must help the young to embrace complexity while finding their way toward understanding.

A thirst for simple answers to complex problems?

Sadly, we see a drift toward mentalsoftness and a preference for entertainment over hard news.

Pandering to these wishes are a host of politicians, pundits, talk show hosts, commentators and modern day self-appointed soothsayers quick to offer up nonsense — simple minded solutions to complicated problems.

Problems with illegal immigrants? Build a wall.

Problems with schools? Test students much more frequently.

Problems balancing the budget? Borrow, print money and give tax cuts.

Sadly, decades of school research rituals reinforce a cut-and-paste mentality that leaves citizens poorly equipped to think for themselves or manage complexity. As stressed in the first chapter, schools must end these rituals and engage students in research that requires that they construct answers and make up their own minds. They must teach the young to embrace complexity.

Entertaining complexity

Following the publication of *Learning to Question to Wonder to Learn*, I began asking audiences to ponder a diagram much like the one on the next page. Comprehending shares center stage with curiosity, pondering, wondering and a dozen information literacies.

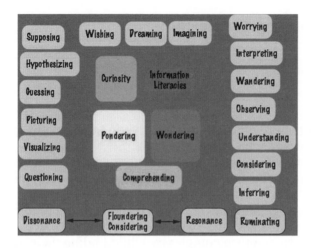

Most times I have asked audiences what they found most surprising about the collection of words, and their responses have varied considerably. Worrying and wandering drew many votes along with floundering.

During this time, I came to appreciate the value of "growing" a diagram like this one, adding new words and concepts as they came to mind. Frequently I added a word because it seemed delightful or illuminating, but at other times I found myself adding ideas that seemed tangential or even discordant. The diagram began to stretch out of shape in surprising ways, and this stretching was productive rather than disturbing.

Instead of compressing or simplifying the array as a cook might reduce a sauce, I came to see that extension and elaboration were valuable. Value resulted from juxtaposition and dissonance. The odd mixture was provocative and productive.

I first tasted this approach playing with clusters of words on the *Visual Thesaurus*. Give it a try. http://visualthesaurus.com

Search for "question" on the *Visual Thesaurus* and surprising words appear like "marriage proposal" (as in "pop the question.") The same happens with many other words. Our understanding is expanded rather than compressed.

Later in this chapter we will explore how schools can encourage students to embrace complexity and manage ambiguity while developing advanced comprehension skills that will help them pierce through the fog of modern and post modern life. We will consider the relationship between comprehending on the one hand and curiosity, pondering, wondering and a dozen information literacies on the other hand.

Pandering to simple-mindedness as an industry

We are offered dozens of books devoted to Dummies . . .

- Taxes for Dummies
- Leadership for Dummies
- The Bible for Dummies
- Voting for Dummies
- Parenting for Dummies

Visit the Web Site at http://www.dummies.com/

This cultural drift is ominous. If we have so little tolerance for what Michael Leunig calls "the difficult truth," (see his poem *Verity* at http://fno.org/may01/verity.html) then how can we hope to wrestle with reality?

As Beth Orton sings . . .
"Reality never lives up to all that it used to be."
Later she adds, "The best part of life, it seemed, was a dream."
Lyrics from "Best Bit" on her album *Pass in Time* (2003)

Comprehending across a dozen information literacies

If schools must equip the young with the skills to understand their world, then they will prepare them for delving into different categories of information - each of which represents an essential literacy:

Text Literacy	Numerical Literacy	Cultural Literacy
Natural Literacy		Social Literacy
Artistic Literacy	**Media Literacy**	Ethical Literacy
Kinesthetic Literacy	Environmental	Emotional Literacy
Visual Literacy	Literacy	Scientific Literacy

When meanings are not directly stated and must be inferred, the comprehension challenge is dramatically exacerbated. Students who are accustomed to finding explicitly stated meaning may not know

29

how to manage when significance is implicit and obscured or simply unavailable. Latent truths require digging and synthesis rather than simple scooping. They actually may not even exist until pieced together by the thinker.

While schools may not have devoted much attention to synthesis in the past, it has become the crux of our challenge as educators in this century. We have reached a stage that requires the construction or making of meaning from clues, data and bits of evidence that will only make sense when we have learned to handle them like the pieces of a jigsaw puzzle.

That puzzling depends on a potent mix of wondering, pondering, wandering and considering. The simple convergent thinking of previous decades does not suffice, especially when the questions or issues at hand are more mysterious than puzzling.

Puzzles often have right answers and their pieces may actually fit together nicely, while the mysteries may prove more elusive.

Googling for the truth

Both young and old come to trust *Google* or *Wikipedia* to provide them with the truths they need, whether it be the lyrics to a song, the best treatment for prostate cancer or an inexpensive hotel in Shanghai.

Often the results of such searches prove useful and even illuminating, but the limitations of this prospecting may not be evident to all. Some who turn to *Wikipedia* do so without realizing that the authors of articles may be very biased or poorly informed.

Google offers the option of clicking on "I'm feeling lucky."

They thereby invite us to surrender to their judgment as to which Web site best serves the question asked. "I'm feeling lazy," is a more accurate wording for such surrender. How reliable is their judgment? Judge for yourself. Read the article, "Feeling Lucky? Surrendering to Pundits, Wizards and Mindbots." http://fno.org/oct06/lucky.html

Wondering, pondering, wandering and considering

For the young to make their own meanings, they need to be able to shift back and forth across the four operations of wondering, pondering, wandering and considering. Each has its own role and its own time, but sometimes they may operate concurrently.

Under the old topical approach to school research, students were asked to do little more than scoop and smush.

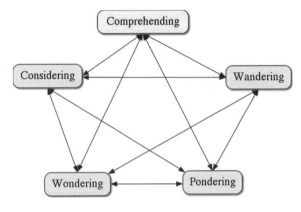

It was all about finding information rather than building an answer or solving a mystery. Little comprehension was involved. Under this approach a student might learn when George Washington was born and what battles he fought, but it was rare that schools might ask questions of import like those below requiring some real wondering, pondering, wandering and considering:

1. In what ways was the life remarkable?
2. In what ways was the life despicable?
3. In what ways was the life admirable?
4. What human qualities were most influential in shaping the way this person lived and influenced his or her times?
5. Which quality or trait proved most troubling and difficult?
6. Which quality or trait was most beneficial?
7. Did this person make any major mistakes or bad decisions? If so, what were they and how would you have chosen and acted differently if you were in their shoes?
8. What are the two or three most important lessons you or any other young person might learn from the way this person lived?
9. Some people say you can judge the quality of a person's life by the enemies they make. Do you think this is true of your person's life? Explain why or why not.
10. An older person or mentor is often very important in shaping the lives of gifted people by providing guidance and encouragement. To what extent was this true of your person? Explain.
11. Many people act out of a "code" or a set of beliefs which dictate choices. It may be religion or politics or a personal philosophy. To what extent did your person act by a code or act independently of any set of beliefs? Were there times when the code was challenged and impossible to follow?
12. What do you think it means to be a hero? Was your person a "hero?" Why? Why not? How is a hero different from a celebrity?

Such questions of import quoted from the *Biography Maker* cannot be answered with simple scooping and smushing. They are in line with the demanding comprehension questions asked by the NAEP reading test.

The *Biography Maker* is at http://fno.org/bio/biomaker.htm

Wandering and stumbling to meaning

When students have experienced only the scooping and smushing model, they have little grasp of what it means to explore a question of import. Basic to this kind of research is the realization that because they do not know what they do not know, they must be careful not to organize their research around what little they do know, as the limits of their understanding may block them from uncovering much that is novel or surprising.

Questions of import, being puzzles or mysteries, almost always require that the researcher muck around a bit, digging into the raw information like a detective seeking clues.

When it comes to biographies, students are usually accustomed to reading secondary sources and have little awareness of how to delve into correspondence, diaries, news accounts or other sources that might cast light on the person's character and provide the kind of evidence that could support them in forming their own opinions.

Relying upon secondary sources is especially dangerous with historical figures like Thomas Jefferson or Captain James Cook because many of them have been turned into icons and heroes. The biographies may have "Photoshopped" the hero, leaving out unflattering stories, atrocities and major character flaws.

For an example of Photoshopping, take a look at the Dove video called "Evolution" at http://www.campaignforrealbeauty.com/flat4. asp?id=6909. Chapter Eight of this book, "Photoshopping Reality: Journalistic Ethics in an Age of Virtual Truth," explores this challenge in depth.

A vivid example of such distorted and unsubstantiated material can be found on the Web site of the Canadian National Library:

> Many believe that James Cook was the greatest ocean explorer ever to have lived, and that he mapped more of the world than any other person. It cannot be denied that he combined great qualities of seamanship, leadership and navigational skill. He mapped many parts of the Pacific, including New Zealand, Hawaii and the west coast of North

America. He also set an example of how to treat a ship and
its crew.
http://www.collectionscanada.gc.ca/explorers/kids/
h3-1710-e.html

While Cook was remarkable in many respects, this paragraph
glosses over his human failings, his errors of judgment, his erratic
patterns of discipline and the damage he did to the native people he
encountered. A complex man is quickly turned into icon.

The diaries and letters of historical figures can sometimes provide
a very different view of these people than what we might find in an
encyclopedia or other secondary sources, but sometimes encyclopedias
do a surprisingly good job, as is the case with "Thomas Jefferson —
American in Paris" an *Encyclopædia Britannica* article at http://www.
britannica.com/eb/article-61878/Thomas-Jefferson

> But the Paris years were important to Jefferson for personal
> reasons and are important to biographers and historians for
> the new light they shed on his famously elusive personality.
> The dominant pattern would seem to be the capacity to live
> comfortably with contradiction. For example, he immersed
> himself wholeheartedly in the art, architecture, wine, and
> food of Parisian society but warned all prospective Ameri-
> can tourists to remain in America so as to avoid the avarice,
> luxury, and sheer sinfulness of European fleshpots.

We must teach the young to wander back and forth between
secondary and primary sources, learning from the opinions of others
while testing them against the evidence.

After reading the *Encyclopædia Britannica* article, for example,
we might read Thomas Jefferson's love letters to Maria Cosway. No
matter how carefully we review the evidence, however, his romantic
relationships will probably always remain a mystery, whether it be
with Maria or his relationship with a household slave, Sally Hennings,
who may have given birth to several of his children.

In contrast to Jefferson, George Washington's daily journal entries
tell us almost nothing about his thoughts and very little detail about his
activities. Famous people vary greatly in terms of the tracks they leave
behind, and primary sources can prove as misleading and distorted as
secondary sources.

It takes some wandering about to learn about these people. If
the student finds GW's diaries disappointing, she or he must wonder

where else to look. Wondering leads to renewed wandering that in turn inspires considering and pondering, all of these in service to the goal of understanding or comprehending the character of the figure being investigated.

Introducing wandering as a research approach

Because many students will arrive in class with nothing much more than the scoop and smush model, teachers must acquaint them with the wandering model mentioned above by conducting what amount to information "field trips" — taking them on productive wanderings that lead to insight, discoveries and an occasional "Aha!" Knowing that this type of learning may be entirely new to students, teachers must provide what amounts to a field guide to wandering about.

That field guide must be adapted to match the kinds of data available for various disciplines, with social studies and science each having their own types of information landscapes and challenges, but one good example of such resources might be "Using Primary Sources in the Classroom" available at the American Memory section of the Library of Congress.

http://lcweb2.loc.gov/learn/lessons/primary.html

This notion of purposeful wandering occurred to me before reading *A Field Guide to Getting Lost* by Rebecca Solnit, but her book is a wonderful extension of the concept and a delightful read.

Chapter 5 - Managing the Poverty of Abundance

The Web offers so much information that students may find themselves effectively starved and frustrated, lost and overwhelmed. The masses of documents and images confront them with what amounts to the poverty of abundance. Unless they can find the image or the document they need, the heaps of material may serve to obscure rather than reveal or illuminate. Schools must equip the young with powerful search strategies so they can manage this challenge, ferreting out material that is pertinent and reliable.

A simple search in *Google* for "China" returns more than a billion Web pages on most days.

In April, 2009, *Google* offered 702,000,000 pages for China. Two years ago, it was 1,200,000,000 pages. Note the round number and many zeroes. Are they really counting? How many of those pages are about china plates? Back in 2004, they reported more than 2 billion Web pages.

Too much information can act like fog, obscuring rather than illuminating.

The importance of search logic

Even though *Google* and other search engines join with many software companies in supplying simple search mechanisms that supposedly require little thought, it turns out that pinpointing pertinent and reliable information usually requires the use of search logic and most often works best when students (and adults) employ advanced search techniques.

There is no searching for dummies!

Sadly, many students and teachers go right to the simple version of *Google* rather than the advanced version, thereby losing the power to find the needle in the haystack or diamond in the rough. In one session with 100 teachers working on laptops, I first asked them to do a *Google* search for "China," then asked how many used the advanced version of *Google*. Only three raised their hands — teacher librarians all three.

What is search logic?

Search logic is the thoughtful combination of search terms - much like a boxer's combinations of punches - that allows the researcher to eliminate irrelevant information and focus on the most promising sources.

Take a look at your options at *Google's Advanced Search* http://www.google.com/advanced_search

Google's Advanced Search offers a series of search boxes, each of which can help you to focus your search.

Looking for climate information to help decide which of six Chinese cities has the least rain?

- "with the exact phrase" allows you to enter "average annual rainfall" so the 702,000,000 pages for China becomes just 27,000, still too many for most of us to read, but great progress.
- "without the words" allows you to enter "plates, hotels, food" so the 27,000 pages for China becomes just 9,560, still too many for most of us to read, but more great progress.
- Language allows you to select "English" so the 9,560 pages for China becomes 8,800.
- Domain allows you to enter ".gov.cn" for Web pages from the Chinese government so the 8,800 become just 107.

We have come closer to the needle in the haystack!

How do teachers and students learn search logic?

Given the flood of information that arrived with the Net some ten years ago, one would expect that search logic would be a high priority for professional development, yet it is difficult to find a program that has made this goal a focus.

A *Google* search for "search logic," "professional development" and "school" turned up just 282 Web pages. We quickly move from the poverty of abundance to slim pickings. There are very few useful resources or proven models to guide schools.

Every school district should check its professional development offerings to make sure that each teacher has a chance to learn search logic using activities like "Searching for the Grail: Learning How to Find Quality Information" available at http://fnopress.com/search.html.

Most search engines offer tips and resources to strengthen search skills. *Google*'s can be found at http://www.google.com/intl/en/help/refinesearch.html.

Students should also be taught search logic and the use of advanced search engines so they can zero in on sources that will actually shed light on their questions. This instruction can be provided by teacher-librarians as part of an explicit series of lessons in information literacy skills but should also be incorporated in subject areas where research is frequently expected.

Figuring out how to use each site

There are hundreds of rich Web sites that offer huge collections of documents, artifacts and images. Unfortunately, many of these are not organized to support efficient searching and there are few recipes or menus that can be transferred from site to site.

The researching student must figure out how to make each of these work. This will usually require some reading of site-specific help pages as well as some trial-and-error efforts.

Example One - Corbis.com — When you can find what you want

In contrast to the next sites, Corbis.com, which sells images to advertising agencies, has constructed an image collection with meta tags and key words that allow you to search through tens of thousands of images and find just what you want without wasting time.

Search for "kittens drinking milk" and you get thumbnails of kittens drinking milk.

Search for "old woman playing poker" you get NO RESULTS.

Search for "old woman playing cards" you get 17 images.

Search for "crying young student" you get some very sad pictures.

If all Web sites offering images, documents and artifacts had databases constructed with the skill of Corbis, we and our students would suffer much less from the poverty of abundance.

Example Two - PictureAustralia.org

This is a resource rich site that allows a researcher to search across the collections of dozens of museums and other sites:

> Search for people, places and events in the collections of libraries, museums, galleries, archives, universities and other cultural agencies, in Australia and abroad — all at

the same time. View the originals on the member agency web sites and order quality prints at your leisure.

The beauty of this site — its capacity to search the collections of so many institutions — is also its biggest disadvantage. The results of searches can be massive.

If you use the simple search for "Captain Cook" and search all the collections, you will get 611 images, many of which seem irrelevant because so many places now use his name.

It takes skill and ingenuity to separate the pertinent from the extraneous.

If you go to Advanced Search and search for "Captain James Cook" with quotation marks, you get just 94 images, most of which are portraits.

The basic search only looks at title, creator, subject, date or place. Advanced search looks at all fields. Thus, basic search turns up 48 images of kitten while advanced search turns up 81 images.

Example Three - Library of Congress *American Memory Collection*

This is another resource rich site that can prove frustrating, especially since its digital collection is much smaller than its main collection. Many online visitors assume that they will be able to find just about everything they might ever want to know about the American experience at this site, but the digital collection is far from comprehensive, and the list of topics not covered is much longer than the list that is covered.

Sometimes it works better to browse the list of subject headings or collections to find out what is there before conducting a more targeted search.

When you compare what the Library offers in the way of resources devoted to Culture & Folklife (34 collections), the list is surprisingly brief. http://memory.loc.gov/ammem/browse/ListSome.php?category=Culture,+Folklife

While some of these 34 collections are real treasures — The North American Indian, a collection of photographs by Edward S. Curtis being just one example — the list does not come close to capturing the American experience. http://memory.loc.gov/ammem/award98/ienhtml/curthome.html

Chapter 6 - Digging Deep

A superficial look at any important question or issue is unlikely to unearth much of value or produce much enduring understanding, yet the culture often seems content with the glib, the facile and the ill-considered. No wonder Bernard Madoff was able to separate so many good folks from their money. Few bothered to do due diligence. There was enough face validity (and charm) to satisfy many.

The *New Oxford American Dictionary* lays out the concept *superficial* with some elegance:

> *Superficial* — existing or occurring at or on the surface :
> the building suffered only superficial damage.
> - situated or occurring on the skin or immediately beneath it: the superficial muscle groups.
> - appearing to be true or real only until examined more closely: the resemblance between the breeds is superficial.
> - not thorough, deep, or complete; cursory : he had only the most superficial knowledge of foreign countries.
> - not having or showing any depth of character or understanding: perhaps I was a superficial person.

This book will keep returning to issues of quality, suggesting that all students must learn to think clearly and deeply, but trends are headed elsewhere. For some years now I have suggested that we might be facing an "Age of Glib."

> With the shift toward electronic media and information, the challenge of knowing and comprehending is complicated by a movement toward superficial and plastic coverage. Deep thinking, deep reading and deep commentary are replaced in many quarters by sound bites, mind bites, eye candy and mind candy.

> Does this Mind Candy Kafe presage the arrival of an Age of Glib?

Are we entering a phase during which we will confuse the accumulation of massive piles of information with the attainment of Truth?
"The Mind Candy Kafe: Replacing Truth with Placebo."
http://www.fno.org/jun98/kafe.html

The antidote to this cultural drift is to equip all students with the skills and the inclination to probe below the surface and dig until they find what is needed to reach a solid level of understanding.

A matter of attitude

How do we nurture dogged determination?

It will not be easy, given the ease with which students can scoop up information from *Google, Wikipedia* and other such sources. What is the pay off for such a struggle? Who really cares? What is wrong with viewing Princess Diana and Margaret Thatcher as icons?

Perhaps we can win the enthusiasm of students by increasing their awareness of knock-offs, counterfeits and frauds. When research is viewed as a mystery waiting to be solved, the dynamics shift away from the tedium of scooping. The student becomes a detective on the trail of a missing person, a missing truth, a missing solution.

Missing pieces and missing clues go along well with the concept of *negative space* employed by artists to describe the space around an object. Traditional school research focused on what was known. True research will focus on both the known and the unknown. For some strong visual examples of negative space, read "Negative Space in a Painting," by Marion Boddy-Evans, at About.com. http://painting.about.com/od/paintingforbeginners/ss/negativespace.htm

In order to develop student appreciation of depth, schools must devote attention to the concept in an explicit manner, incorporating the concept into class discussions, assessment rubrics and school goals.

A matter of skill

This book outlines the skills required to dig deeply enough to find substantial evidence and support insight. By developing questioning skills, students learn to go to the heart of the matter instead of skimming along the surface. Chapter Twelve, "Showing Students How to Challenge Assumptions," Chapter Thirteen, "The Evidence Gap," and Chapter Twenty-One, "Reading Between Digital Lines" all address such skills.

Chapter 7 - After NCLB
Back to the Basics of Inquiry and Comprehension

After an extended and painful educational Dark Age, the USA may be about to return to the crucial learning required by a healthy democracy: inquiry-based with a focus on comprehension and a curriculum richly and broadly defined. The lean offerings and narrow focus of recent years will hopefully be replaced by what will seem in comparison like a feast as the factory style learning, the script-driven instruction and the oppressive emphasis upon frequent mediocre testing all shudder to a stop and the nation awakens to brighter possibilities.

The recent obsession with reading and math in the USA fostered by the narrow minded NCLB law has turned attention away from serious problems with comprehension, understanding and thinking skills that have plagued us for decades when measured by the NAEP tests — the National Assessment of Educational Progress. Note the article, "Adolescent Literacy: Putting the Crisis in Context" by Jacobs, Vicki A, in the Spring 2008 issue of *The Harvard Educational Review*.

When both political parties joined forces in Congress to authorize NCLB, they imposed a national regime of test-driven learning on this generation of children. It was the worst educational policy decision of two centuries. The damage done is huge and the failures many. Fortunately, the insanity of this kind of narrowly defined education has finally dawned on many people. The next decade promises a return to a full-bodied kind of schooling — one that engages students in building the comprehension skills that have declined in recent years. We can also look for policies that encourage students to remain in school rather than drop out in frustration.

Why comprehension?

Few citizens noticed the decline in thinking across the nation as the NCLB juggernaut rolled across the nation, sweeping away decades of progressive practice in the name of so-called research-based instruction. If you teach students nothing but scoring well on dumbed down tests, it should surprise few that the results will be disappointing.

Those states like Connecticut that set high standards and employed challenging tests were urged to switch to cheaper tests that measured less worthy goals. The net result of NCLB has been the dumbing down of a nation's schools and schooling as the policy stressed punishment, shame and public humiliation as motivators.

Children, especially the poor and the economically disadvantaged, have been robbed of their educational birthright as schools served double and triple doses of math and reading while eliminating recess, social studies, science and thinking.

Comprehension across a dozen categories of information requires that students be able to make inferences and think for themselves. They cannot get by with patterns memorized after thousands of pages of worksheets. Drill and kill learning prepares them for low-level testing, but not for the test of life or the kinds of items on the National Assessment of Educational Progress. Those states that bent to the pressures of NCLB dumbed down their tests and ultimately their students as proficiency levels rose on the state tests but declined on NAEP. Our nation's children and schools deserve much better treatment.

Evidence of decline and failure

While Washington made dubious claims of progress due to NCLB, the evidence spoke quite eloquently of failure. The NAEP results were pretty much flat while other indicators were quite alarming. Propaganda replaced authentic reporting of results.

Education Week reported a growing problem
April 10, 2007
"Skills Gap on State, Federal Tests Grows, Study Finds"
By Lynn Olson

The study by Policy Analysis for California Education, a nonprofit research group based at the University of California, Berkeley, was released here during the annual meeting of the American Educational Research Association. The researchers compiled state and federal testing results for the period 1992 to 2006 from 12 states: Arkansas, California, Illinois, Iowa, Kentucky, Massachusetts, Nebraska, New Jersey, North Carolina, Oklahoma, Texas, and Washington.

In all but two states—Arkansas and Massachusetts—the disparity between the share of students proficient on state reading tests and on NAEP, a congressionally mandated program that tests a representative sample of students in every state, grew or remained the same from 2002 to 2006. A similar widening occurred between state and federal gauges of math performance in eight of 12 states.

What comes next?

With Congress arguing over (and delaying) the re-authorization of No Child Left Behind, there is some evidence that the original enthusiasm for NCLB has been replaced by disillusionment and disaffection. A strong current running through those feelings is resentment toward Washington's strong arm approach to state governments and the imposition of policies by distant bureaucrats. The overly zealous focus on testing as a silver bullet has produced none of the results predicted by its proponents but many other consequences that have dismayed a wide array of parents and educators. With a new Congress and a new President, we can expect that there will be a backing away from the heavy handedness of the first NCLB and a move toward restoring state control of educational policy.

Chapter 8 - Photoshopping Reality
Journalistic Ethics in an Age of Virtual Truth

The phrase "journalistic ethics" may become an oxymoron as the media play for this society the role the circus played for Rome — entertaining rather than educating. Jumbo shrimp. Journalistic ethics. Airplane food. Oxymoronic.

How do young people learn what is happening in their world when the attention of the news cameras swings after just two days of coverage from North Korea's nuclear test to the airplane crash of a Yankee player in Manhattan?

The typical news day is dominated by stories that are sensational or entertaining rather than illuminating. Consequently, our view of the world and its events is routinely and daily narrowed, darkened and distorted.

A news day at CNN — April 10, 2009

These stories were featured at *http://cnn.com:*

- Ship nears Kenya without captured captain 45 min
- Gunman kills 1, wounds 3 at Dutch cafe
- WWLTV: Woman, 2 kids slain near New Orleans
- Weekend smackdown: Miley vs. mall cop
- Ticker: Muslim issues dominate agenda
- Car hits churchgoers after Good Friday service
- Mom, infant killed by Tennessee tornado
- Man charged with murder in pitcher's death
- 'Catastrophic failure' caused North Sea crash
- Clothes-stealing cat victimizes neighbors Video
- Phelps? mom calls bong flap 'gossip' Video
- Actor Fox stays positive in face of disease Video
- iReport.com: Easter basket, foot bath, shoes
- Woman sees home invaded on webcam
- Cops yank woman out of car after chase Video
- Harrelson: I mistook photog for zombie
- CNN Wire: Protesters force postponement...

Content analysis

What portion of the stories listed above was made up of basic news stories and what portion was devoted to entertainment, sports, celebrity lives, scandals, weird events and stories with shock value?

Photoshopping

Thanks to new technologies, we can adjust our public images, hiding pounds, ankle bracelets and other burdens of truth that might conflict with the message we hope to sell.

"No pain, no gain!" is an outmoded saying when weight loss can be achieved with the click of a mouse.

The same can be done with news events. It is called "perception management." If a war is not going well, a government may control the flow of information to "spin" things positively. But those who live by the sword of spin may also fall on it as the media can turn on the spin meisters if carnage and bad news pulls a better audience than cheery messages of progress.

Photoshopping reality?

- When the media focus on celebrity break-ups and scandals rather than the still suffering citizens of New Orleans, they are photoshopping reality.
- When government officials make false claims about progress, they are photoshopping reality.
- When critics focus on the negative and ignore the positive, they are photoshopping reality.
- When pharmaceuticals push the use of pills while down playing side effects or hiding knowledge of dangers, they are photoshopping reality.
- When corporations like Enron cook their books, they are photoshopping reality.
- When states fail to count minority students and use easy tests to create the impression of progress for NCLB, they are photoshopping reality.
- When publishers produce history texts that avoid controversy and leave out embarrassing moments, they are photoshopping reality.

Implications for schools

It is evident that schools must equip all students with a dozen or more literacies, not just the simple and narrow numerical and text literacies now fashionable with right wing politicians and their minions. The well being of any democracy is threatened when young people can pass through school with little awareness of the issues properly addressed within the scope of media literacy. If they are not equipped to consider news critically and thoughtfully, how will they vote or plan their lives?

Media literacy deserves prominent placement in district curriculum documents, especially in language arts, social studies and science. It is all too often ignored or shoved aside onto a "library skills" track.

Types of literacy

Text Literacy	Numerical Literacy	Cultural Literacy
Natural Literacy	**Media Literacy**	Social Literacy
Artistic Literacy		Ethical Literacy
Kinesthetic Literacy	Environmental Literacy	Emotional Literacy
Visual Literacy		Scientific Literacy

Far too much attention in schools is devoted to the gathering of information without showing students how to assess the reliability of sources. If we expect them to be media savvy, we must identify a repertoire of critical thinking skills that would sustain such savvy and then offer learning experiences that gradually nurture mastery of those skills as the students pass through our programs.

Program strategies

Students should be given ample opportunities to look critically at news coverage, sometimes in English class, sometimes in social studies, sometimes in math and sometimes in science.

For example, a class of high school English or social studies students might look at the USA edition of CNN and compare it with the international edition. Beginning with http://cnn.com, they could critique the focus of the USA edition and then open a second window for the international edition at http://edition.cnn.com/.

The juxtaposition of the two should heighten their awareness of how news media focus their attentions.

Math and science students might consider news coverage of a concept such as global warming by going to http://news.google.com/ and searching for articles on that topic. On April 10, 2009, they would have found many articles, the first ten with titles listed below:

- Rice-duck farming can mitigate global warming
- Easter in crisis
- NASA uses satellites to fight wildfires
- Breaking the silence about Spring
- Public Hearing On Off-Shore Oil Drilling and Protection of ...
- Obama discussing the use of pollutants to stop warming
- Gordon Brown shows how green he really is
- Uri Gordon: Why the G20 can't deliver
- Install a Fireplace in any Room of the House
- Three point-guard lineup works — once

Such articles offer an opportunity to critique the math or science behind the headlines. Students might explore the following questions:

1. How sound is the mathematical (or scientific) thinking or research mentioned in this article?
2. Who are the authors and chief researchers cited in the article and what do we know about their credentials, their credibility and their potential conflicts of interest?
3. Is the information in this article trustworthy?
4. How does the information and thinking in this article compare with information in more scholarly articles we have been reading?
5. Is any truth lost or sacrificed when the media reports on topics of mathematical or scientific import?
6. What techniques, if any, were used to push a point of view?
7. How could this article be improved?
8. Of all the articles you critiqued, which are the least credible and which are the most credible?
9. Where can you find the most reliable information when trying to understand complex issues that influence the quality of life and have math or science as a focus?

Another strategy to emphasize media literacy with social

studies, science and math students is to have them visit *Numeracy in the News*, a Web site created by Australian teachers who offer hundreds of newspaper articles that are all accompanied by discussion questions calling for students to evaluate the findings reported and whether they are justified by the data collected.
http://www.mercurynie.com.au/mathguys/mercury.htm

Consider the following questions as you read the articles.

- Does the article present enough information about how the study was carried out to reach a conclusion?
- Are you confident about the inference drawn?
- Was the sample size appropriate in your opinion?
- What questions would you ask the investigator and/or the reporter about the article?

In addition to these suggestions, there are many other activities that have been developed by good teachers and those committed to making media literacy an important part of the school experience. Some can be found by visiting the Web sites listed below:

- **The Center for Media Literacy** (CML) — a nonprofit educational organization that provides leadership, public education, professional development and educational resources nationally. http://www.medialit.org
- **PBS TeacherSource** — Media Literacy - Related PBS Sites and Programs. http://www.pbs.org/teachersource/ media_lit/related_sites.shtm
- **The Henry J. Kaiser Family Foundation** — Media Literacy - Key Facts - http://www.kff.org/entmedia/ Media-Literacy.cfm
- **Montana Media Literacy Performance Standards** http://www.opi.mt.gov/pdf/Standards/ ContStds-Media%20Lit.pdf
- **The Media Awareness Network** — A non-profit organization promoting media education. Their web site, one of the largest educational web sites in Canada, provides expansive media and web literacy resources for parents, educators and community leaders. Great lesson plans for teachers at http://www.media-awareness.ca
- **The Media Literacy Clearinghouse** http://www.frankwbaker.com/

Chapter 9 - Teaching Media Literacy in an Age of Wikilobbying, Spin, and Tabloid Journalism

The previous chapter laid the groundwork for this one, demonstrating that many of the sources available to our students will distort information in one way or another. Finding or even approaching evidence that is credible and reliable is quite a challenge. Verification is an essential skill and checking the veracity of sources is a major component of media literacy. Understanding production techniques and their impact is essential.

What do we mean by literacies?

Each literacy involves the capacity to interpret or make sense of information within a particular category or mode. This concept is also explored in Chapter Twenty-Two.

In a democratic society, we would expect that all children will learn to understand information within and across a dozen categories at the least, but there is a disturbing global trend to focus on just two of these: numerical literacy and text literacy.

The list of literacies in the previous chapter places *Media Literacy* in the center because so much of what comes at students is generated by the media outlets.

Text Literacy	Numerical Literacy	Cultural Literacy
Natural Literacy		Social Literacy
Artistic Literacy	**Media Literacy**	Ethical Literacy
Kinesthetic Literacy	Environmental Literacy	Emotional Literacy
Visual Literacy		Scientific Literacy

How is media literacy different?

Media Literacy is mainly concerned with the news, advertising and entertainment that flows from TV screens, computer screens, cell phone screens, magazines, newspapers and radio stations, all of which are hoping to shape our understanding of the world as well as our consumer and our citizenship behaviors.

In many respects, the critical thinking skills required to achieve some degree of media literacy are much like those that serve well for the other literacies. Students must develop the capacity to step back from each message and ask penetrating questions:

- Where is this information coming from, and what are they hoping I will think, feel and do?
- What techniques are being used to sway my thinking, my feelings and my actions?
- What are the subtexts?
- What can I believe?
- What can I trust?
- What is real?
- What is fake?
- What is misleading?
- What should I check?
- What should I challenge?

Several decades back, thinkers warned of these times:

When the real is no longer what it used to be, nostalgia assumes its full meaning. There is a proliferation of myths of origin and signs of reality; of second-hand truth, objectivity and authenticity.

Jean Baudrillard, *Selected Writings*, ed. Mark Poster (Stanford; Stanford University Press, 1988), pp.166-184.

In these times of facsimiles and digital impressions, simulacra will often replace or stand in for what was real.

In 1954, Ernest Hemingway stated in an interview with Robert Manning that, "To invent out of knowledge means to produce inventions that are true. Every man should have a built-in automatic crap detector operating inside him."

In 1971, Neil Postman and Charles Weingartner popularized the phrase "crap detector" in *Teaching as a Subversive Activity*.

The history of the human group is that it has been a continuing struggle against the veneration of "crap." Our intellectual history is a chronicle of anguish and suffering of men who tried to help their contemporaries see that some part of their fondest beliefs were misconceptions, faulty assumptions, superstitions or even down right lies.

> We have in mind a new education that would set out to cultivate ... experts at "crap detecting" ... people who have been educated to recognize change, to be sensitive to problems caused by change, and who have the motivation and courage to sound alarms when entropy accelerates to a dangerous degree. (pp. 3-4)

If crap detectors were needed then, they are most probably even more crucial these days.

Wikiality - when Wikipedia defines reality

According to Alexa.Com, *Wikipedia* had a traffic rank of seventh as this book was being written, meaning that it attracted more visitors than all but six other Web sites in the world. In contrast, venerable, so-called professionally crafted encyclopedias such as *Encyclopedia Britannica* ranked 2,952. This data was gathered in March of 2009.

With such heavy traffic and dominance of Google search results, *Wikipedia* has the power to influence the thinking of the entire world. As Stephen Colbert pointed out on the Monday, July 31, 2006 edition of *The Colbert Report* news program, Wikiality represents "A reality where, if enough people agree with a notion, it becomes the truth."

> Any user can change any entry and if enough other users agree with them, it becomes true. ... If only the entire body of human knowledge worked this way. And it can, thanks to tonight's 'Word': Wikiality. Now I'm no fan of reality, and I'm no fan of encyclopedias. I've said it before: Who is Britannica to tell me that George Washington had slaves? If I want to say he didn't, that's my right. And now, thanks to *Wikipedia*, it's also a fact. We should apply these principles to all information. All we need to do is convince a majority of people that some factoid is true.
> —Stephen Colbert, "The Word," *The Colbert Report*, July 31, 2006

What makes the popularity and dominance of *Wikipedia* doubly ironic is the questionable nature of the professional encyclopedias all along. By their very nature, these lengthy tomes have always distorted

53

truth in various ways, either by oversimplifying, shortchanging or misrepresenting the subjects at hand. For decades they were weak on women and minorities. They have always reported the facts of cities without speaking much to the soul of the town. While editors crow about experts at the helm, much of the writing is actually done by junior staff members.

It is fascinating to lay articles from *Wikipedia* and *Encyclopedia Britannica* side by side. Take Paris, for example. It is difficult to rule on which source does a better job of capturing the spirit and character of the city. Both articles are quite rich in detail and information. One cannot automatically dismiss *Wikipedia* as an inferior source in many instances.

But *Wikipedia* raises new and different issues as articles remain open to changes from just about anybody who wants to edit without having to sign in or identify themselves. Those articles that touch on controversial subjects or reflect on the fortunes of individuals and corporations are especially vulnerable to tampering and vandalism, as Steven Colbert points out with considerable humor:

> This is the essence of wikilobbying, when money determines *Wikipedia* entries, reality has become a commodity. IBM could throw some of their money at perception and make their product 'objectively better,' then Microsoft can just fire their cash cannons back and we're off to the races. This is the essence of wikilobbying.
>
> http://www.colbertnation.com/the-colbert-report-videos/ 81454/january-29-2007/the-word---wikilobbying?videoId=81454

A number of researchers have revealed tampering with articles by various groups. They traced the IP numbers of those who have made the edits. The *New York Times* reported on these efforts in "Seeing Corporate Fingerprints in *Wikipedia Edits*," August 19, 2007:

> Dozens of similar examples of insider editing came to light last week through WikiScanner, a new Web site that traces the source of millions of changes to Wikipedia, the popular online encyclopedia that anyone can edit.
>
> The site, wikiscanner.virgil.gr, created by a computer science graduate student, cross-references an edited entry on

Wikipedia with the owner of the computer network where the change originated, using the Internet protocol address of the editor's network.

http://www.nytimes.com/2007/08/19/technology/ 19wikipedia.html?ex=1189828800&en=1c6d59cafa5f5e84 &ei=5070

If you go to wikiscanner.virgil.gr and enter the name of an organization or corporation such as Microsoft, the following results appear:

New Search **WikiScanner - select organizations**

Search results for query "Microsoft"			
Edits Ahoy!			
organization name	domain	location	# edits
Microsoft Corp	-	Redmond, Washington	37748
Microsoft Corp	tigspecialty.com	Redmond, Washington	9656
Microsoft Corp	msn.net	Medford, Oregon	6373

Thirty-seven thousand edits is a very large number of edits. In contrast, there were only a few hundred from IP addresses at Wal-Mart.

Some of the edits, as the *New York Times* article points out, are swiftly reversed by the guardian or wizard responsible for each article. This was the case with the article on Sarah Palin that suddenly experienced dozens of positive edits the night before her selection as VP candidate. Most of those changes were reversed and the article was locked down to prevent further adjustments. Note the article in the *International Herald Tribune* at http://www.iht.com/articles/2008/09/01/ business/link.php "Editing — and re-editing — Sarah Palin's *Wikipedia* entry."

At the same time we have reason to be concerned about *Wikipedia*, it does a better job on some subjects than conventional encyclopedias, so it would be foolish to ban it (as China did) or ignore it. At times the contributions of amateur editors are richer in detail and feeling than is the case with conventional encyclopedias. Biographies and cities may benefit from this richness while they might also suffer from bias, as fans of historical figures may leave out important stories of failures or misjudgments. We may end up with sanitized versions of their lives.

When I first read the *Wikipedia* article on Matthew Flinders, a British ship captain important for his early exploration of the

Australian coast, the article was silent on a number of his missteps and errors of judgment. I stepped in and added the incidents to set the record straight. Those changes remain a year later. When *Wikipedia* lives up to its claims, it can serve quite well as source. However, as outlined in Chapter Two, *Wikipedia* suffers from a number of problems:

- Quality, reliability and language
- Bias and balance
- Currency

Since our students will join the adult world in making frequent use of this source, the wise strategy is to equip them with the crap detecting skills mentioned at the beginning of this article. We would hope they would apply such skills to all information sources, whether reading about a ship captain like Matthew Flinders in *Wikipedia*, the *Enclyclopedia Britannica* or in one of several biographies written about him. His biographers saw him very differently from each other, and the *Enclyclopedia Britannica* hardly notices him at all, giving him just 170 words and one image, in contrast to 2200 words and many illustrations in *Wikipedia*. The supposedly professional source distorts his place in history along with that of Australia by downplaying his importance on the world stage — an incredible instance of bias against a historical figure and a continent — bias through neglect and inattention.

In this case and in many others, the power of dozens of enthusiasts can develop information that is sometimes superior to that offered by the so-called experts. In a frightening sense, all information has become somewhat suspect, and we must teach students to assess its reliability without trusting too much in its source.

Spin

To convey information or cast another person's remarks or actions in a biased or slanted way so as to favorably influence public opinion; information provided in such a fashion. http://www.wordspy.com

Complicating our search for truth and understanding is the prevalence of *spin* in the culture. Almost nothing is what it seems to be since nearly everyone in the media is carefully shaping perception. There is a Web site devoted to "the Spin of the Week." http://www.gnn.tv/articles/3951/Spin_of_the_Week

Banks, manufacturers, universities, government agencies and politicians are all busy trying to shift how events are viewed. It matters less and less what actually happened. What matters is perception. The society is filled with masters of deception who can turn catastrophe and shameful incompetence into poetry and redemption.

The Wizard of Oz has handed down his many tricks and tactics so that "smoke and mirrors" are now basic communication tools. One highly effective media design firm even calls itself "Smoke and Mirrors" and has a Web site by that name at http://www.smoke-mirrors.com/smweb/. They were running an ad for "Save the Children" as this chapter was written in March of 2009 that suggested that those who are busy saving banks should try saving children. Powerful work.

The supposedly unreliable *Wikipedia* does well with this term:

> Smoke and mirrors is a metaphor for a deceptive,
> fraudulent or insubstantial explanation or description. The
> source of the name is based on magicians' illusions, where
> magicians make objects appear or disappear by extending
> or retracting mirrors amid a confusing burst of smoke.
> The expression may have a connotation of virtuosity or
> cleverness in carrying out such a deception.

In contrast, *Enclyclopedia Britannica* has no entry for the term.

This spinning of tall tales puts students and all citizens in a difficult position, since we rarely hear or see what is really happening. We are shown what the spinners want us to see. Media literacy can serve as an antidote, as heightened awareness and skill puts students on the alert to pierce through the manipulation and distortion.

An early stage of developing media literacy might be labelled "awareness." Just as previous chapters argued that students should be made aware of Photoshopping, spinning should be high on the list of cultural realities that all students learn about as they pass through middle school into high school and college. Awareness is only the beginning of a process that will prove quite demanding.

Tabloid journalism

On most days there are hundreds of news events around the globe, but current news programming will usually ignore most of these events, pandering instead to the public's appetite for scandal, gossip, disaster and celebrity escapades. While the newsroom was once insulated from the business operations of most media outlets, that is no longer the case in most companies as the drive for ratings has pushed what used to be called "journalistic ethics" to the side.

These are difficult times for journalists as consolidation and cost factors have combined to reduce reporting staff and shift focus away from news to what is often entertainment.

Los Angeles Times lays off 10% of its editorial staff
The newspaper cuts 75 employees as advertising revenue continues to decline.
By Roger Vincent — October 28, 2008

Facing falling revenue in a stalling economy, *The Los Angeles Times* on Monday laid off 75 editorial employees, part of a 200-person reduction that began last week.
http://articles.latimes.com/2008/oct/28/business/fi-times28

Because there is little market and little support for serious journalism, many important stories go unreported and unnoticed. Likewise, serious journalists go unhired. A pretty face may serve the "news reading" function better than a battle-weary veteran of real stories.

This lack of attention to serious news events is clearly a sin of omission, but the public is complicit, as a media outlet offering two dozen true news stories would probably find its ratings dropping dramatically as folks switched to other channels.

Various studies have documented a decline in the level of interest shown by most groups for hard news. Leading the decline would be young people in their teens and 20s.

For extensive data on the media habits of various generations go to http://www.frankwbaker.com/mediause.htm

The New York Times has reported a steady decline in the reading of the more serious newspapers.

Newspaper Circulation Continues to Decline Rapidly
By Richard Pérez-peña Published: October 27, 2008

The long decline in newspaper circulation over the years continues to accelerate, with sales in the spring and summer falling almost 5 percent from the previous year, figures released on Monday show, deepening the financial strain on the industry.

The drop occurred nearly across the board during the six

months that ended Sept. 30; weekday circulation for the largest metropolitan dailies fell anywhere from 1.9 percent for *The Washington Post*, to 13.6 percent for *The Atlanta Journal-Constitution*, compared with the period a year earlier.
http://www.nytimes.com/2008/10/28/business/media/28circ.html

These market pressures have changed the ways that news is gathered, undermining the commitment to investigatory reporting and shifting reliance to news feeds and services so there has been a homogenization and consolidation of reportage.

At the same time, the push for ratings has led to significant crossover to entertainment by the more serious media outlets as news programs have been shunted aside for news magazines and anchors have been chosen for their capacity to stir folks up rather than calm them down and educate them. It is now fashionable to deliver the news on TV stations with personalities taking center stage rather than information and facts.

One example of this trend would be "chat news" — a program that gathers some pundits together to chat about the news of the day or week. Inexpensive to produce, such programs amount to opinion swapping and speculation rather than serious reporting.

Can the Google Generation google well?

In contrast to the many inflated claims about the savvy of young people put forth by those who speak of digital natives and savants, a number of studies have emerged painting a less encouraging picture of the information skills of young people.

One study conducted by the University College London (UCL) CIBER group (2008) "Information behaviour of the researcher of the future," found that the facility with which young people could navigate masked somed disturbing weaknesses in their ability to fathom what they were breezing past. Just a few of their findings are listed below, but the rest deserve some serious attention:

- The information literacy of young people, has not improved with the widening access to technology: in fact, their apparent facility with computers disguises some worrying problems
- Internet research shows that the speed of young people's

web searching means that little time is spent in evaluating information, either for relevance, accuracy or authority
- Young people have a poor understanding of their information needs and thus find it difficult to develop effective search strategies
- As a result, they exhibit a strong preference for expressing themselves in natural language rather than analysing which key words might be more effective
- Faced with a long list of search hits, young people find it difficult to assess the relevance of the materials presented and often print off pages with no more than a perfunctory glance at them

University College London. CIBER Briefing paper; 9. http://www.jisc.ac.uk/media/documents/programmes/reppres/gg_final_keynote_11012008.pdf

A New Zealand study also found weaknesses in the information literacy skills of students that contrasted with their inflated sense of confidence:

- Students are more "confident" than "competent" in the "information literacy" skills in their first year of tertiary education.
- Secondary-school-based "information literacy" skills did not transfer well to the tertiary environment.
- Secondary school teachers were more confident than their tertiary colleagues that their students would have appropriate "information literacy" skills to transfer to the tertiary environment.
- There is a discrepancy between secondary and tertiary understandings of students tertiary "information literacy" requirements
- The library is low on the list of where students would go for assistance if they felt they needed "information literacy" support to complete their studies.

Source: "Fishing with the new net" Information skills transfer from secondary to tertiary environment: a presentation on the results of an Otago, New Zealand secondary and tertiary Information Literacy survey. Shields, G., & Bennett, D. http://akoaotearoa.ac.nz/project-register/531

Such studies underline the importance of teaching students media literacy because they substantiate a serious gap between expectations and performance. As studies show young people and their elders shifting their habits and adopting new media sources for their information needs, the pertinence of media literacy becomes evident.

Who is responsible for teaching media literacy?

If schools accept the responsibility of teaching media literacy, they need also to consider the delivery options. In some places, there is a fairly narrow reliance upon the teacher librarian to deliver on this goal as part of a larger commitment to an umbrella term — information literacies. In other schools, the English and language arts department may shoulder prime responsibility. Other departments such as social studies, science and art may also accept a significant role, especially when the school recognizes the importance of engaging students in the construction of media as well as the de-constructing of media.

The capacity to produce media-rich communications seems to be an essential component of any program meant to equip the young with an understanding of the game, thereby inoculating them to some extent against the perception management, spin and Photoshopping so prevalent in the society.

The Center for Media Literacy is a superb resource to guide a school's thinking about smart ways to incorporate media literacy into its classroms. Their vision and their mission are very much in concert with the theme of this chapter:

Vision

The Center for Media Literacy (CML) is dedicated to a new vision of literacy for the 21st Century: the ability to communicate competently in all media forms as well as to access, understand, analyze, evaluate and participate with powerful images, words and sounds that make up our contemporary mass media culture. Indeed, we believe these skills of media literacy are essential for both children and adults as individuals and as citizens of a democratic society.

Mission

Our mission is to help children and adults prepare for living and learning in a global media culture by translating media literacy research and theory into practical information, training and educational tools for teachers and youth leaders, parents and caregivers of children.

http://www.medialit.org/

Chapter 10 - Gaining Attention
in the Laptop Classroom

Over the years we have seen little time or funding devoted to professional development (PD) that might equip teachers with the skills required to make smart and productive use of new technologies. What little PD we have seen was devoted mainly to the learning of software and how to operate the latest toys. We have too often placed carts before horses. To use these tools in support of original thinking — the focus of this book — requires a substantial investment in professional development that would emphasize effective teaching and learning strategies.

Managing learning in the laptop classroom requires considerable savvy. This chapter explores the dimensions and characteristics of such classrooms, identifies the chief challenges facing teachers in such space, and suggests moves, tactics, strategies, tricks and scaffolding designed to optimize student learning.

Pedagogy does matter, but it has been pretty much ignored during the design of PD offerings. Note the article "Pedagogy Does Matter!" at http://www.fno.org/sept03/pedagogy.html

A *Google* search for "laptop classroom" and "pedagogy" turns up few articles with much to offer. The one exception is a study by Ewa McGrail at Georgia State University, "Laptop Technology and Pedagogy in the English Language Arts Classroom," *Journal of Technology and Teacher Education*, 2007. 59-85.

McGrail's abstract echoes the themes of this chapter:

The English Language Arts teachers in this qualitative study reported somewhat negative outcomes in social and material spaces in the context of laptop technology in their classrooms. These outcomes included: (a) social isolation, (b) limited communication with a teacher or peers, and (c) offtask behavior. In an attempt to uncover the reasons for these rather negative results, the researcher analyzed these teachers' classroom environments and instructional engagements with laptop technology, since these practices are believed to be reflective of these teachers' current beliefs about instruction and technology's role in it. Some of the reasons the researcher uncovered were: (a) limited physical space, (b) cumbersome furniture, (c) poor technology infrastructure and (d) the largely instrumental use of technology in numerous learning engagements.

A successful lesson results from mindful devotion to the following elements, each of which will be addressed later in this chapter:

1. Lesson objectives, content and design
2. Classroom landscape
3. Classroom culture
4. Teaching moves, strategies, tactics and procedures
5. Equipment
6. Assessment procedures

An effective teacher knows how to orchestrate these elements to match the characteristics of the students, adapting to each group as the lesson proceeds. It is difficult to design a lesson in advance because teaching and learning involve interaction and chemistry. The teacher should be alert to adjustments required by this interaction. What works well with a second period class may fail when tried with a different group at the end of the day.

1. Lesson objectives, content and design

The presence of laptops in a classroom should not dictate the nature of lesson design. That would be allowing the tool to determine the learning. Sound instruction calls for clarity about goals prior to the selection of equipment. Sadly, in many schools, the tools have become defining elements when they should not be. A laptop school? A laptop classroom?

Would it not be more encouraging to witness the spread of thinking schools and questioning classrooms along the lines set out in this book? Our goal is to grow young thinkers. This is not about laptopping children. It is about educating them.

We should not do technology for the sake of technology or go digital simply because it is fashionable. We begin by establishing learning goals, and then we identify the activities most likely to carry the day.

Example: A high school English teacher developing the media awareness of students makes the critical thinking and comprehension skills embedded in media literacy the focus of the unit's lessons. Eager for students to apply these skills across a range of media, the teacher shifts back and forth from print to television to online blogs, magazines and news Web sites like CNN as mentioned in Chapter Eight — Photoshopping Reality.

At times during this unit on media literacy, all books and all laptops will be shut. The teacher may display an ad up on the classroom's screen where all can focus.

> "What techniques have been used in this ad to sway your emotions and influence your thinking?"

The teacher has chosen a technique that brings the class together with eyes forward. Perhaps they could have all seen the same ad on their separate laptops, but classroom interaction would have been quite different. The teacher has their attention — an aspect of classroom learning that is inadequately considered in the literature on new technologies in schools.

The ad being viewed by the class comes from the Dove Foundation. Mentioned earlier in this book, "Dove Evolution" shows a young woman being Photoshopped for a billboard, ostensibly raising issues about the impact of the beauty industry on young women's self images. http://www.campaignforrealbeauty.com

The teacher downloaded the file from the Web at home, knowing that the school's network would perform poorly when trying to show such streaming video. At the same time, the decision to show the video up front was determined by a desire for attention at the outset of the lesson.

> "Now open your laptops and work as pairs to critique and analyze this same ad, frame by frame. I want one of you making a list of techniques on one of the laptops while you use the second laptop to watch the video I passed around earlier on thumb drives."

As the lesson and the unit proceeds, the teacher is clear about purpose and intentions. The class will shift in and out of various equipment configurations according to the benefits of each. At times students will work solo on separate laptops. At times they will team in groups of six, sharing just two laptops. At other times, the class will sit in one big circle and share personal reactions while all equipment is cooling.

2. Classroom landscape

Landscape is rarely discussed in schools. In some, the desks, the tables and the chairs rarely move. Visitors can walk down the halls of many high schools and find nearly all the classrooms organized with rows of desks facing the front. At the front sits a teacher desk and a lectern as well as a presentation table that might hold equipment such as an overhead projector or a video projector and document camera. Sometimes equipment hangs from the ceiling or peers out from a corner.

This design has been popular for many decades and serves classroom lecturing fairly well, but effective teachers will ask students to move furniture around to meet lesson objectives, since some arrangements serve particular learning goals better than others. Constructivist learning activities, for example, thrive when furniture is not arranged in rows as it is for didactic lessons.

Example: To illustrate this landscape issue, consider a seventh grade social studies teacher asking students to debate the wisdom of offshore drilling.

The class begins with all seats facing forward. The teacher points to a chart on the white board showing fastest times for eight different furniture arrangements in this class:

1. Lecture
2. Discussion
3. Debate
4. Solo Research
5. Pair Research
6. Team Invention
7. Team Dialogue/Interpretation
8. Model City Council

She holds up a stopwatch.

"OK, now, get into debate format when I give the signal; all who oppose the congressional offshore drilling moratorium on this side of the room and those who support it on this side of the room."

It takes the students less than a minute to change the room to match the learning activity. They can now glance across the room and concentrate on their opponents and the ideas at stake. The teacher may still direct the flow, but attention has been shifted intentionally away from the front of the room to the other students.

On some days the teacher may allow students to keep laptops open to consult research done in previous classes or to take notes. There may be a ground rule that blocks browsing during the debate. The teacher may have developed class norms that distinguish between productive and acceptable browsing in contrast with diversionary wandering that would dilute the attention being paid to the debate. On some days all laptops will be shut.

When networked computers first came to schools, most were desktop machines that did not move easily. The flexible approach to classroom landscape outlined above was very difficult with such equipment and many schools built computer labs will most of the equipment bolted to the floor. Many of those labs had computers lined up in rows facing the front of the room. Teachers in such classrooms had little choice but to adjust to a single reality, except that they could change the front of the room to the back by stepping behind the students where they would be able to see their screens.

A major message of this chapter is the importance of teachers being able to see the students' screens when they are working on challenges and tasks assigned by the teacher. While there is now software available to allow such monitoring from the teacher's computer, most teachers must rely upon a technique we will call eyeballing. Certain furniture arrangements are more supportive of eyeballing than others, and thanks to the arrival of wireless laptops, movement and flexibility are now real options in many schools.

Even though movement and flexibility are options, many teachers have worked through decades of arranging rows to face the front. Those habits are sometimes deepseated and hard to shed.

During the past ten years, I have successfully employed the arrangement shown in the diagram on the next page to handle class sizes of 50—175 working hands-on as pairs or trios.

This arrangement of tables and laptops was designed so that I would be able to see every screen when class members were working on information challenges. This eyeballing would help me know how to pace the lesson and would allow me to pay attention to those in need.

In this situation, when I need the full attention of the class for 5—10 minutes, I will ask them to swivel around and join me. They

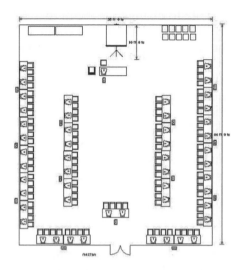

turn their eyes forward on me and the screen. Sometimes I am initiating a group discussion, in which case they will swivel around and face each other in one large C-shaped grouping. Sometimes I will invite them to turn with laptops on laps because they have important work just completed that they might need to share. Other times I will ask participants to turn their backs on the laptops because they might prove a distraction.

Most times when I hold these sessions, several teachers will enter the room and move the laptops around so they are sitting on the opposite side of the tables from the side I intended. Accustomed to facing the front of a classroom, they are most comfortable sitting facing the front of the classroom with a laptop between me and them.

This is a teachable moment, as it offers an opportunity to mention classroom landscape and provide a brief explanation of why I have set up the room in the way I have. Even after this explanation, some are still skeptical. As the day proceeds, we hold a number of discussions of landscape, monitoring, eyeballing and attention. By the afternoon, some of the skeptics have seen the light and embraced the strategy. Others are tenacious, sticking with what they have always known.

During the day, there are times when participants are asked to gather with six chairs and two laptops. At other times they are urged to go anywhere in the building for fifteen minutes to work on a challenge. The configuration that greets them in the morning morphs, shifting and adjusting as the activities change.

A teacher who ignores landscape in laptop classrooms is likely to encounter difficulties with classroom management. A useful Web site outlining some of these issues can be found at
http://www.asbindia.info:8081/drupal/Tech@ASB/?q=node/81

3. Classroom culture

Culture is yet another topic neglected by those who believe equipment will transform classrooms and learning. Success with a room full of laptops requires attention to attitudes and norms. The behaviors of students will either support or undermine the learning depending upon the classroom culture.

If a teacher has five classes, it is likely that each will be quite different from the others in terms of attitudes and behaviors. These can be developed by a savvy teacher who understands culture. If a group starts the year as an uncooperative, untrustworthy mob, the teacher acculturates them. Failure to address attitudes and norms would normally lead to chaos. If the group starts off with more positive norms, the teacher takes them to ever more advanced levels. A good teacher invests in organizational development.

What are the most important aspects of classroom culture when it comes to laptop classrooms?

• **Attention** — There will be many times when the class must shift gears from independent work to focus upon the teacher who may be giving instructions or engaging the class in a discussion. Relinquishing the track pad and establishing eye contact must become a well conditioned, reliable habit even though the temptations of the laptop are many.

• **Collaboration** —There are times when students should work in teams to share ideas, interpret data and come up with solutions to problems or challenges. Success requires scaffolding of cooperative skills such as active listening and taking turns. The teacher introduces a dozen skills early in the year and gives the class lots of practice along with rubrics so they know what effective behavior looks like.

• **Independence** — The capacity to work steadily, alone or in teams, without much adult prodding or supervision is both a matter of attitude and skill. The teacher must scaffold the skills required to perform whatever tasks have been assigned but must also cultivate the spirit that is required. Many students may be used to prior years of passivity - classrooms where teachers provided the answers and required little of the students. The teacher must make it clear to students that independence is required and will be assessed on a continuous basis.

• **Resourcefulness** — When students are engaged in making answers rather than simply absorbing them, they must wrestle with complexity and struggle with difficult questions. They cannot cut and paste. They will not be enjoying spoon feeding. They must learn to feed themselves. The thesaurus suggests the following related qualities: entrepreneurial, imaginative, ingenious, inventive, creative;

quick-witted, clever, bright, sharp, sharp-witted; enthusiastic, dynamic, proactive, ambitious, energetic; bold, daring, courageous, adventurous. As with independence and the other elements listed below, the teacher must make it clear to students that resourcefulness is required and will be assessed on a continuous basis.

• **Versatility** — In a laptop classroom committed to exploration and thinking, there will be false starts, frustration and difficulty. Students must develop the deftness to shift approaches until something works.

• **Responsibility** — Not all students walk into class prepared to work through activities in a manner that respects the norms outlined by the teacher. Some will test those expectations, violate the norms and see if they can get away with unacceptable behaviors. These students might waste time on entertainment Web sites, for example, when the teacher is not looking. They may engage in instant messaging. Their compliance with norms may rely entirely on enforcement rather than personal commitment. The teacher must identify those students who lack such commitment and confer with them through the year, nudging them toward responsibility. At the same time, the teacher must supply the monitoring and classroom discipline required to make sure such students do not lapse. It is unwise to presume independence before it has been verified.

• **Persistence** — Struggling toward understanding is a bit counter-cultural at a time when many think that all questions can be swiftly answered with a *Google* search or a quick read of *Wikipedia*. In order to meet the expectations stated by the American School Librarians and ISTE, students must do more than gather information when they research. The AASL sets the expectation that students will develop good new ideas, not just scoop and smush, copy and paste. This kind of learning will take time, be frustrating and require staying power.

4. Teaching moves, strategies, tactics and procedures

A laptop classroom requires quite a repertoire of moves, strategies, tactics and procedures to make sure learning occurs. Sadly, professional development in few districts has supplied much assistance to teachers who wonder about the management of such classrooms.

There are hundreds if not thousands of specific moves and tactics that can help to make a teacher successful in this kind of classroom. This chapter cannot list all of them, but it will offer some concrete examples.

Example 1: An effective teacher will instruct students to move the furniture around as mentioned earlier to match the landscape to the activity.

Example 2: An effective teacher will move about the room when students are working on laptop challenges, checking, monitoring, assisting and intervening or supporting as necessary.

Example 3: An effective teacher will scan the room with frequency to figure out where to move next but will also use eyeballing and eye contact to make some of those movements less necessary as many students respond well to such gestures.

Example 4: An effective teacher will announce a coming shift of activity with enough time so that students can wrap up what they are doing. "In two minutes I will be asking you to turn around and join me for the next activity."

Example 5: An effective teacher will ask students to "shut laptop lids" when laptops would not prove useful. Sometimes a single laptop shared by a pair or trio works better than one-on-one. Sometimes a group discussion is best served by cooling laptops.

Example 6: An effective teacher will call for complete attention prior to demonstrating a series of instructions. "I need 100 percent eye contact now."

Example 7: An effective teacher will hold many conferences with individual students and teams, guiding, suggesting, encouraging and supporting their work in various ways.

Example 8: An effective teacher will devote much energy to assessment, most of it informal, asking how things are going and what needs to be adjusted.

Districts with an interest in laptop classrooms (whether they be one-on-one or shared laptops) would be wise to offer a 2-3 year professional development offering that would focus on equipping all teachers with these moves, tactics and strategies. This program would offer guided practice and mini-teaching as well as support groups so that teachers can acquire the skills by doing rather than mere listening and watching. Teachers move from theory into practice thanks to learning communities.

5. Equipment

The most effective laptop classrooms will provide a full range of technologies rather than surrender to simplistic laptop definitions. In some schools, the desire to give every student a personal laptop leads to some very bad choices.

1. Depending on the funds available, one-on-one may force the purchase of underpowered, heavy laptops with the wrong features, speed and performance.

2. Depending on the funds available, one-on-one may leave no money for a classroom projector, a decent screen, blackout shades, a document camera, an interactive whiteboard or appropriate furniture.

3. Depending on the capacity of the school network, a move to one-on-one may overwhelm that network so that students experience performance disappointments such as slow access to Web sites.

The most effective laptop classrooms will view laptops as just one element within a complex array of tools that extends to mundane items such as blackout curtains and chairs. Good planning orchestrates all of these elements so as to optimize learning, but grandiose visions sometimes subvert such planning. One can visit laptop classrooms where every student has a computer but the room is so awash with sunlight that the projector cannot be used. There are also many classrooms where the chairs and tables are better suited to a didactic classroom of the 1950s. Putting all one's eggs into a laptop basket leaves no money for swivel chairs or trapezoidal tables. While rarely noted, a classroom with shared laptops is also a laptop classroom.

The choice of equipment should closely match the learning activities intended for that classroom space. Sadly, to question the wisdom of one-on-one computing is like pointing out the Emperor's nakedness. Many classrooms end up missing important items thanks to the one-on-one myth. Authentic discussion of alternatives is often blocked by the egos of so-called visionaries who are intent on imposing their ill-considered dreams on others. Often these visionaries work outside of schools and presume to know best how to improve an enterprise about which they are fundamentally ignorant. Big bucks do not always translate into effective policy, as can be seen with the ill-fated small schools effort launched by the Bill and Melinda Gates Foundation. Note article "Good Doubt and Bad Doubt" at http://questioning.org/jan07/doubt.html.

6. Assessment procedures

As mentioned earlier, the effective teacher will devote a good deal of energy to gauging the effectiveness of lessons, strategies and activities, employing a range of techniques that include eyeballing and conferencing as well as rubrics developed by the students themselves. To the extent that these assessment techniques become shared classroom practices, the program can deepen and improve.

There have been few convincing studies of the impact of laptop programs upon student learning, in part, perhaps, because visionaries rarely tolerate such verification. Most studies list doubts such as Dunleavy and Heinecke, "What added value does a 1:1 student-to-laptop ratio bring to technology-supported teaching and learning?" in the *Journal of Computer Assisted Learning*:

> Results indicated that online research, productivity tools, drill and practice, and eCommunications were the most frequent uses of computers in the 1:1 classroom. Moreover, the 1:1 classroom provided potentially transformative added value to these uses while simultaneously presenting unique management challenges to the teacher. In addition, the presence of 1:1 laptops did not automatically add value and their high financial costs underscore the need to provide teachers with high-quality professional development to ensure effective teaching. In order to create effective learning environments, teachers need opportunities to learn what instruction and assessment practices, curricular resources and classroom management skills work best in a 1:1 student-to-networked laptop classroom setting.
> http://pkp.sfu.ca/ojs/demo/present/index.php/jce/article/view/171/56

The preferred assessment model of the three monkeys is a recipe for failure.

Who was Joan of Arc?

There is the Joan of Arc we read about in *Wikipedia* and the *Encyclopedia Britannica*.

National heroine of France
. . . a peasant girl who, believing that she was acting under divine guidance, led the French army in a momentous victory at Orléans that repulsed an English attempt to conquer France during the Hundred Years' War. Captured a year afterward, Joan was burned by the English and their French collaborators as a heretic. She became the greatest national heroine of her compatriots.
Encyclopedia Britannica
http://www.britannica.com/EBchecked/topic/304220/
Saint-Joan-of-Arc

Many of these articles turn Joan into an icon, a bit like Princess Diana. She has been Disneyfied — transformed into a caricature, as when the Hunchback of Notre Dame becomes cute and Pocohantas becomes a dazzling cover girl. Note images at http://www.meredith. edu/nativeam/pocahontas_4.jpg.

When we read her letters, an image of her character emerges:

Wherever I come across your troops in France, if they are not willing to obey I shall make them leave, willing or not; and if they are willing to obey, I will have mercy upon them. She comes on behalf of the King of Heaven — an eye for an eye - to drive you out of France. And the Maiden promises and guarantees that she will cause such a great clash of arms that not for a thousand years has there been another one so great in France.
Source: Joan of Arc's 'Letter to the English' (22 March 1429) translated by Allen Williamson
http://archive.joan-of-arc.org/joanofarc_letter_Mar1429.html

The tone of her letter to the English seems a bit imperious when you read the entire piece. If one did not note that she was illiterate and could not write, one might assume the words were hers and use them to substantiate a critical assessment. In fact, according to the translator, Allen Williamson, the letter was dictated to a cleric and Joan later denied having dictated certain portions or using the third person. The tone we note may have been the voice of the scribe, not the voice of Joan. It will remain a mystery, as with much of history.

When we turn to statues, posters and paintings, the task of verifying the portrayal is also quite challenging.

Photos of Joan of Arc statues on this page and the next page were shot in Paris by Jamie McKenzie

It is difficult to figure out who the real Joan was, even if we take the time to read the transcripts from her two trials and her correspondence. But it is even more difficult to find an image of Joan that stands up to scrutiny and might resemble the real Joan.

As far as we can tell, there is no image available today that was created during her time by someone who actually saw her and painted her from life.

Unfortunately, many of the images so freely available to students online may distort the person's character in a number of ways, whether they be learning about Joan of Arc, George Washington or Matthew

Flinders. In a cut-and-paste culture, students may be a bit too quick to capture an image of the person they are studying, snatching it up for a report without considering whether it is authentic.

According to the creator of jeannedarc. info, Søren Bie, we have only one image of Joan created during her lifetime and that artist never actually saw her. http://jeanne-darc.dk/img_jeanne/onlyknowndrawing_big.jpg

His Web site offers an extensive view of Joan of Arc images created over the past several centuries. The main index page for these images can be found at http://jeanne-darc. dk/p_art_image/art.html An especially vivid collection by one artist can be found at http://jeanne-darc.dk/p_art_image/0_art/boutet_de_monvel.html

The artist, Boutet de Monvel had very strong feelings about Joan. While vivid, art of this kind may exaggerate her feats, glorify her personality and contribute to a mythology somewhat in conflict with reality and history.

Taking images at face value

In concert with the rest of this book, we would hope that students will learn to question the validity of historical images as a source of information about figures in history. We teach them not to take images at face value.

An activity that works well to develop this questioning engages the students in collecting images about a figure from history like Joan from *Google Images*. They may paste their collection like the one on the side into a word processing document or a mind mapping program such as *Inspiration*™ or *Smart Ideas*™ .

Instructions:
1. Go to *Google Images* and find at least ten images of your person from history. Keep track of the locations of these images on the Net.
2. Once you have your collection, try to rank them in terms of historical accuracy, placing the most accurate on top and the least accurate at the bottom.
3. While ranking the images, develop a case for why you are ranking them in this way. What is it about the image that lends it more or less credibility than the others?

In the case of Joan, there is some testimony in the transcripts of her trials that can be used to assess the accuracy of the images. Depending on the age of the students involved, the teacher can either point them to specific pages or simply point them to the long, scrolling list of documents and challenge them to find pertinent testimony.

In the transcript from "Saint Joan of Arc's Trial of Nullification" located at http://www.stjoan-center.com/Trials/#nullification, they can read testimony that describes her appearance.

The teacher sends them to the testimony of JEAN DE NOVEL-EMPORT, Knight, called Jean de Metz at http://www.stjoan-center. com/Trials/null04.html and asks them what they can learn from his words:

> When Jeannette was at Vaucouleurs, I saw her dressed in a red dress, poor and worn.
>
> I asked her if she could make this journey, dressed as she was. She replied that she would willingly take a man's

dress. Then I gave her the dress and equipment of one of
my men. Afterwards, the inhabitants of Vaucouleurs had a
man's dress made for her, with all the necessary requisites.

They rapidly learn that it is difficult to nail down the details of
her appearance in a satisfying manner. Historians have been wrestling
with these issues for years because the comments in the trials were
mostly about her conduct and her religious principles.

Stephen W. Richey devotes several pages to this challenge in his
biography, *Joan of Arc: The Warrior Saint (2003, Westport, Conn:
Praeger.)*. He devotes Appendix C to a review of "Joan's Personal Appearance." He begins his attempt with the following admission:

There is not enough surviving evidence to give us any
certainly as to what Joan looked like.

Students can find Richey's analysis by doing a Google search for
the exact phrase "Joan's Personal Appearance." This will take them to
his book and the Appendix C online.

By engaging students in the search for evidence and following
up with the testimony of an historian searching through the clues, the
teacher gives them a firm grounding in the process of verification so
they will always question the accuracy of images they find rather than
taking them at face value.

Chapter 12 - How to Challenge Assumptions

From the New Zealand Curriculum . . .

"Students who are competent thinkers and problem-solvers actively seek, use, and create knowledge. They reflect on their own learning, draw on personal knowledge and intuitions, ask questions, and **challenge the basis of assumptions and perceptions**."

Challenging the basis of assumptions and perceptions

What must teachers do to equip students with both the skills and the attitude needed so they will routinely examine the underpinnings of proposals, propositions and opinions? Ideas and opinions often circulate unchallenged, as outlined in the next chapter, "The Evidence Gap." They not only circulate, they are quite often embraced as truths or even become policy. When a doctor suggests a pill or a specific treatment, how many patients challenge the assumptions underlying such advice? Often they take the pills and the advice without digging to find out how strongly science supports their decision. How often do they ask about side effects? How often do they ask for data?

"Doubts Grow Over Flu Vaccine in Elderly" — By Brenda Goodman in *The New York Times*
Published: September 1, 2008

The influenza vaccine, which has been strongly recommended for people over 65 for more than four decades, is losing its reputation as an effective way to ward off the virus in the elderly.

"For Widely Used Drug, Question of Usefulness Is Still Lingering" By Alex Berenson in *The New York Times*
Published: September 1, 2008

When the Food and Drug Administration approved a new

type of cholesterol-lowering medicine in 2002, it did so on the basis of a handful of clinical trials covering a total of 3,900 patients. None of the patients took the medicine for more than 12 weeks, and the trials offered no evidence that it had reduced heart attacks or cardiovascular disease, the goal of any cholesterol drug.

Schools must teach the young to ask tough questions so they may challenge the many "pills" being offered them as they make their way through adolescence into adulthood. The term "pills" is being used metaphorically, of course, representing everything from drugs and meds to simple answers to complex problems.

Shall I pop this bitter little pill?
Will this pill do me good?
Should we drill offshore?
Should we build a big fence along the border?
Should we bail out finance companies that speculated wildly?

If schools fail to meet this challenge, the young grow up prone to *mentalsoftness*:

1. Fondness for clichés and clichéd thinking - simple statements that are time worn, familiar and likely to carry
   surface appeal.
2. Reliance upon maxims - truisms, platitudes, banalities and
   hackneyed sayings - to handle demanding, complex situations requiring thought and careful consideration.
3. Appetite for bromides - the quick fix, the easy answer,
   the sugar coated pill, the great escape, the short cut, the
   template, the cheat sheet.
4. Preference for platitudes - near truths, slogans, jingles,
   catch phrases and buzzwords.
5. Vulnerability to propaganda, demagoguery and mass
   movements based on appeals to emotions, fears and
   prejudice.
6. Impatience with thorough and dispassionate analysis.
7. Eagerness to join some crowd or mob or other - wear, do
   and think what is fashionable, cool, hip, fab, or the opposite or whatever . . .
8. Hunger for vivid and dramatic packaging.
9. Fascination with the story, the play, the drama, the show,

the episode and the epic rather than the idea, the question, the argument, the premise, the logic or the substance. We're not talking good stories or song lines here. We're talking pulp fiction.
10. Enchantment with cults, personalities, celebrities, chat, gossip, hype, speculation, buzz and blather.

I coined the term *mentalsoftness* in May, 2000. See "Beyond Information Power" at http:// fno.org/may2000/beyondinfo.html

Picturing an idea

Few people stop to consider what may or may not lie below the surface of an idea or proposal. In working with students, it helps to provide them with visual images of ideas along with their supporting structures. A well formed idea is a bit like an iceberg or a building. There should be plenty of logic, supporting evidence and careful thought below the surface. The idea itself may appear as a simple sentence or two but its value depends upon its foundations much like a cathedral or a sky scraper.

Continuing with the building metaphor, students are encouraged to think of themselves as building inspectors. When buying a house, smart consumers pay for a housing inspection to check all parts of the house before making a final decision.

- Plumbing
- Electrical service
- Appliances
- Paint
- Roof
- Structural soundness
- Asbestos
- Termites, ants and other infestations
- Drainage

The building inspector employs a check list. Students can employ the same approach with ideas, using the checklist provided later in this chapter.

The underlying Structure of an idea

We teach our students that good ideas are bolstered by three main elements:

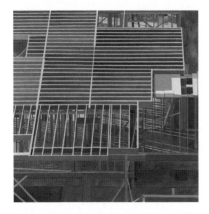

1. Sound Logic
2. Sufficient Evidence
3. Standards

Painting © Sarah McKenzie
originally in color.
Visit her online gallery at
http://sarahmckenzie.com

Students learn to strip away the outer coatings and veneer to see what lies below the surface of the idea. This is what we mean by teaching them to challenge assumptions. First they must identify the assumptions. Then they must see how worthy they are.

- Do the assumptions stand up to demanding questions?
- Are the assumptions based on something more than wishful thinking?   passionate belief?
- Is there abundant evidence to bolster the assumptions?

Example —ß Offshore Drilling in the USA

Prior to the presidential campaign, both Senators McCain and Obama were generally opposed to offshore drilling. As gas prices soared and the American public became increasingly hard pressed at the pump, Senator McCain shifted his position and began arguing for offshore drilling at the same time that President Bush called for drilling, arguing that this act would reduce dependence upon foreign oil and lead to lower prices at the pump. Before long, Senator Obama and the Democratic Party seemed more open to the possibility as public opinion polls showed voters to be enthusiastic about the strategy.

Simply stated, the idea or proposal would read like this . . .

"Drill offshore now to lower dependence upon foreign oil and lower gas prices."

This simple proposition won broad-based acceptance from beleaguered American consumers, most of whom did not take the time to challenge the assumptions lurking behind the proposition. When the Republican President, the Republican Party and the Republican candidate for President all came out in favor of offshore drilling, the Democrats resisted for some weeks but finally caved in under the public pressure, announcing on September 23, 2008 that they would let the 25 year ban on such drilling lapse.

"We want cheap gas and we want it now."

The assumptions?

1. Drilling could start quickly.
2. New gas would flow to the pumps almost immediately.
3. The amount of oil available offshore would make a huge dent in the daily needs of the USA.
4. Adding more oil to the supply is the best way to reduce dependence on foreign oil (in contrast to the development of alternative energy sources, more fuel efficient vehicles and the development of less wasteful consumer behaviors.)
5. The other costs of such drilling (environmental impacts, etc.) are not worth considering and would be inconsequential.
6. If the supply of American oil grew, American demand would not increase.
7. The price of gas is set locally based on American supplies.
8. American gas and energy suppliers have the best interests of the consumers in mind and will do whatever possible to pass along savings and lower prices at the pump.
9. American gas and energy suppliers are American, not global.

10. American culture is a car culture and that culture must not and cannot shift toward mass transit alternatives.
11. Conservation and alternative energy strategies would be slower, more painful and less effective ways to meet the challenge.
12. Large and inefficient cars are essential to the well being of American citizens.
13. Presidential candidates must avoid any mention of pain, suffering or sacrifice if they wish to live in the big White House.
14. Americans cannot rise to challenges that require pain, suffering or sacrifice.

Many of these assumptions are demonstrably false, but they are rarely discussed, considered or challenged. They lurk below the surface. Because they lurk unexamined, they may have an unwarranted influence over decision-making. If the general public knew that many of these assumptions are false, perhaps they would shift away from support.

As noted in Chapter Thirteen, "The Evidence Gap," there are at least two major problems with offshore drilling that stand in contrast to some of the assumptions listed above.

1. New drilling could not start quickly. According to *The Boston Globe*, "It would take at least a decade for oil companies to obtain permits, procure equipment, and do the exploration necessary to get the oil out of the ground, most industry analysts say."

2. New supplies of oil in the USA would not change the price at the pump because the price is set on a global basis. "Suppose the US produced all its oil domestically," said Robert Kaufmann, director of the Center for Energy and Environmental Studies at Boston University. "Do you think oil companies would sell oil to US consumers for one cent less than they could get from French consumers? No. Where oil comes from has no effect on price."
Source: "New offshore drilling not a quick fix, analysts say." June 20, 2008 in *The Boston Globe*, by Lisa Wangsness.

This chapter will not review the rest of the assumptions listed above, for this would then become a chapter about offshore drilling.

The main goal was to illustrate the importance of identifying and challenging the assumptions underlying a policy proposal before jumping on board.

Just How Good is This Idea?

As mentioned above, students can employ the checklist below to determine the soundness of an idea or proposal.

The "Wait Just a Minute!" Idea Evaluator		
Summary of Idea/Proposal:		
		Veracity Rating
Underlying Assumption #1		o Strong o Dubious o Untenable
Explanation for Veracity Rating		
Underlying Assumption #2		o Strong o Dubious o Untenable
Explanation for Veracity Rating		
Underlying Assumption #3		o Strong o Dubious o Untenable
Explanation for Veracity Rating		
Underlying Assumption #4		o Strong o Dubious o Untenable

T*he "Wait Just a Minute!" Idea Evaluator* begins by asking students to identify and evaluate the assumptions underlying the idea. They then proceed to consider the adequacy of the evidence presented.

The form leads the students through a consideration of the logic employed to build the case for the argument or strategies presented.

The image above only shows a portion of the *Idea Evaluator.*

The *"Wait Just a Minute!" Idea Evaluator* can be downloaded at http://questioning.org/oct08/challenging.html for use with students in three formats:

PDF version at http://questioning.org/oct08/wait.pdf
Old MS Word Version at http://questioning.org/oct08/wait.doc
New MS Word Version at http://questioning.org/oct08/wait.docx

Assessing student growth in capacity

The expectations of the NZ Curriculum with regard to testing assumptions can be met with sound instructional strategies and then progress can be assessed using well-tested instruments such as the *Cornell Critical Thinking Test*.

Cornell Critical Thinking Test, Level X (1985), by Robert H. Ennis and Jason Millman. Critical Thinking Press and Software (formerly Midwest Publications), PO Box 448, Pacific Grove, CA 93950. Aimed at Grades 4—14. Multiple-choice, sections on induction, credibility, observation, deduction, and assumption identification.

"What can I make of this?"

Many times our students will encounter claims and statements poorly substantiated by evidence. They are contending with a broad social trend that Stephen Colbert has called "truthiness."

> *Merriam Webster*'s definitions of **truthiness**:
> 1. "truth that comes from the gut, not books. (Stephen Colbert, *Comedy Central*'s "The Colbert Report," October 2005)
> 2. "the quality of preferring concepts or facts one wishes to be true, rather than concepts or facts known to be true" (American Dialect Society, January 2006)

We must develop students' recognition skills so that an alarm goes off when they read or see an unsubstantiated claim, whether that claim be a biographical note, the ad for a drug or the platform of someone running for office. Once the alarm sounds, they must learn to ask for the evidence or data that is missing.

A biographical example

A few years back while looking for Web evidence of Captain James Cook's navigational skill, I encountered a passage on the site of the Canadian National Library, a passage used several times in this book to illustrate distorted history:

> Many believe that Captain James Cook was the greatest ocean explorer ever to have lived, and that he mapped more of the world than any other man. It cannot be denied that he combined great qualities of seamanship, leadership and navigational skill.
>
> He mapped many parts of the Pacific, including New

Zealand, Hawaii and the west coast of North America. He also set an example of how to treat a ship, its crew, and the people he met on his explorations.

Passageways: True Tales of Adventure for Young Explorers at http://www.nlc-bnc.ca/explorers/kids/h3-1710-e.html

This passage has been altered since I lodged a complaint (removal of underlined segment above), but it remains a bold statement of opinion about a complicated man that oversimplifies his record, distorts the truth and relies upon heavy handed language to pressure us into accepting its questionable point of view. There is still an "evidence gap." It is a powerful example of the "Disneyfication of Informatiom" — a trend in media closely associated with truthiness.

Cook routinely took hostages during his three voyages through the Pacific Islands when the native people "borrowed" his equipment or possessions. He preferred the chiefs as hostages and died in Hawaii on the day a chief refused to come along when summoned. Stabbed to death on the beach, Cook was far from a good example of how to treat the native people he encountered. His treatment of his crew is also open to question.

The previous comments are based on information from "The Death of Captain Cook" By Richard P. Aulie at http://www.captaincooksociety.com/ccsu2536.htm and *The Trial of the Cannibal Dog* by Anne Salmond. One wonders if the person who wrote the passage about Cook above actually knew much about his encounters with his crew or the natives he encountered during these voyages.

The term *Disneyfication* refers to information that has been prettified and made more fun, more entertaining and easier to digest than the real version. Information has been glamorized and beautified to make information, history, literature or reality easier to swallow, read or watch.

A pharmaceutical example

In June of 2008, *The New York Times* began a series of articles under the heading, "The Evidence Gap," examining the use of drugs that have been prescribed by thousands of doctors to millions of people despite the lack of complete evidence that these drugs will prolong life or reduce risks from heart attacks. In some cases, these drugs are now thought to increase the risk of cancer.

This series of articles reviews more than six examples of drugs that have been widely adopted without compelling evidence.

The willingness of doctors to prescribe such drugs is a serious issue, especially given the lavish attention and benefits provided to many of them by the drug companies.

But the willingness of patients to accept these drugs without checking doctors' advice is an important aspect of the problem. A patient must understand that the old saying "Buyer beware!" now applies to the medical field.

After a decade of urging seniors to take flu shots, new studies are challenging the design of the original studies that seemed to support the benefits of such shots. Evidently, the seniors in those studies who took shots were generally better off to begin with, were characterized by healthier habits and received care superior to that available to those not taking the shots, so their survival rate may have been explained by factors other than the shots.

A political example

Modern campaigns often rely upon sound bites, mind bytes, eye candy and mind candy to sell candidates, whether they be running on the left or the right. Much campaigning is done with 15- and 30-second campaign ads that compress complicated issues into simple, one line solutions.

It seems that few citizens have the time or the inclination to visit the candidates' Web sites where they would usually find lengthy white papers outlining the candidates' positions on the issues of the day. The thirst for simple, brief answers to complicated problems is met by simplistic solutions that often make little sense. Combined with voters' general dislike of pain and sacrifice, the policy options offered by hopeful candidates rarely involve either.

The issue of soaring gas prices during the first half of 2008 provides a good example. Voters see this as a crucial issue in most countries, but politicians have a difficult time selling conservation as a viable policy option. In the USA presidential election, both major party candidates who previously opposed offshore drilling shifted their stances as polls showed citizens wanting more oil immediately and wanting more drilling right away, even though most experts had testified that it might take 6—8 years or longer before any of this new drilling would bring more oil to market and even though the oil involved might be a "drop in the bucket."

As mentioned in the previous chapter, *The Boston Globe* quoted experts as saying, "about 86 billion barrels of additional oil may lie offshore, according to the US government's Energy Information Administration. Of that amount, about 18 billion barrels are subject to the moratorium. Much of the rest lies in areas that are too expensive to exploit or that oil companies have not yet tapped for technical reasons, fueling the industry's desire for fresh territory."

Furthermore, the article claims, "It would take at least a decade for oil companies to obtain permits, procure equipment, and do the exploration necessary to get the oil out of the ground, most industry analysts say."

Source: "New offshore drilling not a quick fix, analysts say." June 20, 2008 in *The Boston Globe*, by Lisa Wangsness. http://www.boston.com/news/nation/articles/2008/06/20/new_offshore_drilling_not_a_quick_fix_analysts_say/

Numbers do not lie

Obama promised to spend "$150 billion over the next decade in affordable, renewable sources of energy," but the number, while large, is abstract and difficult to grasp. McCain countered with talk about an ambitious project that will, among many other things, "increase the use of wind, tide, solar and natural gas. We will encourage the development and use of flex fuel, hybrid and electric automobiles." He did explain how these things will happen. He did not provide specifics or details, but neither did Obama.

Big numbers like Obama's promise of a $150 billion investment in alternative energy sources are impressive but unnerving. They resonate. They stir. But they also inspire skepticism. Where will he find this money? If we go to his campaign Web site looking for specifics, we do not find them, even if we download the full text of "NEW ENERGY FOR AMERICA" at http://my.barackobama.com/page/content/newenergy It is a big number, for sure, but there is no place in the four page document that clarifies how it will be spent or how it will be funded. There seems to be an evidence gap.

McCain also threw around some big numbers, promising that his plan would help us "stop sending $700 billion a year to countries that don't like us very much." Astute commentators quickly pointed out that the number he used is the total trade deficit, not the trade deficit for oil, and that much of the number was spent on goods produced by so-called friends.

Truthiness

When politicians throw around large numbers or purport to believe in things they have never acted upon, when they wish to be seen green when they are really gray, or when they hope to be seen as well informed when they are really ignorant, they are flirting with what Stephen Colbert has called "truthiness" — a term already mentioned and defined earlier.

When students and citizens learn to poke below the surface claims and demand evidence to substantiate claims and positions, truthiness melts away like the wicked witch in *The Wonderful Wizard of Oz*. Closing the evidence gap is like throwing water on the witch.

Chapter 14 - Bettering: A Synthesis Primer

How do we make life better?
How do we make things better?
How do we better ourselves?
How can we better our town, our company or our team?
How can we better our performance?

Bettering relies on synthesis — a set of change strategies that modify and adapt what is to what it might become. Bettering also requires skill at both analysis and evaluation. This entire book is devoted to originality and the capacity to come up with good new ideas.

Questioning is an essential aspect of effective synthesis. Curiosity did NOT kill the cat!

Questioning is what makes bettering possible. Unfortunately, many teachers report that they have been given very little support or training with regard to synthesis. This chapter will outline professional development strategies to address this deficit.

How can we make this play better?

- What actor can we **substitute** for the lead in order to improve the impact of this play?
- Would Sally's lines to the landlord in Act II come across stronger if her sister joined her in speaking them and their voices were **combined** like a chorus?
- Can we **adapt** the ending so the audience does not leave feeling quite so depressed and disheartened?
- Can we play down or reduce (**minify**) the number of times Andrew seems to be lecturing to the audience?
- How can we involve Sam (the play's major financial backer) in the play (**put to other uses**) in ways that will make him feel valued but end his interference with the director?
- Since the play goes on a bit too long, what sections can we **eliminate** without weakening the impact and the message?
- Ever since we fired the lead, the press coverage of the play has been very negative. How can we turn that around (**reverse**)?

The questions about the play listed on the previous page are an example of how **SCAMPER** works to support synthesis and bettering. Each letter of the word **SCAMPER** stands for a synthesis strategy.

S = substitute
C = combine
A = adapt
M = modify (magnify, minify)
P = put to other uses
E = eliminate
R = reverse

Invented in the 1950s by Osborne (and bettered by Bob Eberle to help business people become more inventive, SCAMPER is just one of many synthesis tools teachers should understand and be able to share with their students.

Construction 6 (Pile) by Sarah McKenzie, © 2008

Synthesis and bettering as constructing

The previous page shows a painting of a home partially construct-ed. It stands as a metaphor for the synthesis and bettering thought process, as a thinker must recombine and modify the elements of the object or process being improved much as an architect and a builder might instruct a carpenter to arrange 2x4s and other pieces of lumber to construct a house.

The real synthesis in home design begins in the planning stages, as an architect plays with thousands (perhaps millions) of possibilities in order to create a house that will satisfy a number of criteria such as cost, comfort, style and environmental responsibility.

By the time the carpenters are hammering nails, there is little room left for synthesis or bettering, as they are usually following a prescriptive plan that leaves little room for modification.

Construction 4 (STEPS), by Sarah McKenzie, © 2008

In June and September of 2001 I published two articles that make good companion pieces for this chapter:

"Building Good New Ideas" http://fno.org/jun01/building.html

That article lists and describes synthesis strategies worth learning.

1. SCAMPER
2. The Questioning Toolkit
3. The Scientific Method
4. von Oech's A Whack on the Side of the Head
5. deBono's Thinking Hats
6. Michalko's ThinkerToys
7. Mental Mapping and Cluster Diagramming

"Building Good New Ideas — Part 2"

http://fno.org/sept01/building2.html

That article employed carpentry in the design of a grape arbor as the central metaphor for the bettering of ideas.

Cooking with Bloom - bettering the omelette

In order to acquaint teachers with the skills associated with synthesis, I have used the following professional development activity with great success. Originally part of a morning I called "Cooking with Bloom," this activity relies upon a right side of the brain, multi-sensory experience to give the audience a memorable experience with synthesis that will stick with them for life. Instead of relying upon dry abstractions, the activity takes advantage of smell, sight, sound and taste.

Step One — Assemble all tools and ingredients.

- Frying pan
- Hot plate
- 3-4 eggs
- Butter
- Red pepper flakes
- Spatula
- Very smelly cheese
- Cooking mirror as used by home economics teachers

Step Two — Explain to your group that the focus of the lesson is upon synthesis, also known as "bettering." You will be cooking an omelette rather badly, and their job is to keep track of the steps on a piece of paper and then suggest ways to improve the cooking process — thereby bettering the omelette.

Step Three — Make sure the frying pan is very hot - then add the smelly cheese and let it melt some until the smell begins to spread through the room and you hear groans. Make effective use of the cooking mirror so the audience can see as well as smell what is happening in your pan.

Step Four — Crack each egg and drop it shells and all into the pan with the bubbling cheese mixture.

Step Five — Take a big chunk of butter and add it next to the eggs and egg shell mixture, commenting that butter is important to keep the omelette from sticking.

Step Six — Empty the entire bottle of red pepper flakes into the pan and stir.

Step Seven — Remind the group that they should be making a list of suggested improvements or changes as the omelette cooks.

Step Eight — Collect suggested strategies up on the screen using a mind-mapping tool such as Inspiration™ and then see if they can be grouped according to the SCAMPER model described at the beginning of this article.

Problem-based and authentic learning: Synthesis on a large scale

When we involve students in suggesting ways to better their community or tackle a nagging social or environmental problem, we are asking them to perform synthesis at the highest level. It is not sufficient for them to study the problem or describe the problem. It is not enough to cut and paste the thinking of the sages. In many of these cases, the sages and the elders have failed, so it is incumbent upon the students to "do a better job" and add value to the action plans of the past.

- Problem-based learning is outlined in some depth at http://fnopress.com/pbl2/toc.html
- Authentic learning for social studies is outlined in some depth at http://fno.org/sept07/soc.html
- Authentic learning for language arts is outlined in some depth at http://fno.org/nov07/lang.html

Hands-on professional development for synthesis

When teachers have a chance to learn synthesis with hands-on activities, their grasp of the approach is stronger than when the concept is just abstract and theoretical. A few years back I created a set of hands-on modules for the NECC conference that addressed this need and the following goals:

Power Learning 2.0
Beyond Interpretation to Synthesis

Expected Outcomes

- Experience firsthand the challenge of moving beyond interpretation and understanding to synthesis - the actual construction of new ideas.
- Learn how to focus classroom investigations around decisions and problems drawn from the community and the global neighborhood. Engage students in making their own meaning (constructivism) from the vast new information landscape that is made readily and rapidly available thanks to new technologies.
- Consider how the role of classroom teacher changes in such a program.
- Witness how we may provide structure and scaffolding to maintain quality and focus.
- Explore how students can learn to build their own new meanings and solutions upon conventional wisdom, contributing fresh thinking to important social issues or scientific challenges.
- Begin to develop a personal repertoire of synthesis skills.
- Identify opportunities to teach to curriculum standards by emphasizing decision making and problem solving.

You will find a description of these activities at
http://fnopress.com/PL2/index.htm

You will find the description of a new hands-on workshop at
http://fnopress.com/SYN/

Chapter 15 - The Mystery Curriculum

Even though the most important questions in life may be to some extent unanswerable, these challenging and often frustrating issues should occupy a prominent position in the curriculum guides of schools that claim to prepare students for life in this new century.

Unfortunately, perhaps because these issues are usually messy and frustrating, many schools sidestep them in favor of learning that is comfortably packaged and topics that are ripe for "plucking."

Few lists of *21st Century Skills* include the ability to manage ambiguity, wrestle with paradox and entertain mystery, but this failure to prepare our students for mystery is a serious lapse. The times are riddled with enigmas and conundrums. If schools adopt a mystery curriculum, students may approach these issues with familiarity and confidence instead of confusion, contempt or panic.

Mysteries vs. puzzles

Put simply, mysteries are thornier and less cooperative than puzzles. There may not be a satisfying answer to a mystery, while most puzzles can be solved, especially those like a jigsaw.

Malcolm Goodwells's article in *The New Yorker* (January 8, 2007) about Enron's mystifying accounting challenges, "Open Secrets," makes this point dramatically. He credits the national-security expert Gregory Treverton:

(Treverton) has famously made a distinction between puzzles and mysteries.

Osama bin Laden's whereabouts are a puzzle. We can't find him because we don't have enough information. The key to the puzzle will probably come from someone close to bin Laden, and until we can find that source bin Laden will remain at large.

The problem of what would happen in Iraq after the toppling of Saddam Hussein was, by contrast, a mystery. It wasn't a question that had a simple, factual answer. Mysteries require judgments and the assessment of uncertainty.

http://www.newyorker.com/reporting/2007/01/08/070108fa_fact

Examples

Even though students around the world may study the events of the Second World War and the horrors of Auschwitz — the extermination of millions of Jews, Gypsies and others — some of the most mysterious and troubling questions embedded in this history may go unexamined in some classrooms.

- How do leaders persuade citizens to go along with such horrors?
- Can genocide be stopped by those caught up in it?
- What kinds of people are willing to participate in the actual slaughter?
- What kinds of people are willing to look the other way and pretend nothing is happening?
- Are we looking the other way right now?
- Where is genocide happening today and what are we doing about it?
- What does it take to stop such horrors from continuing in our own times?
- In what ways can religion play a role in encouraging such horrors? or discouraging them?

- How does the UN define genocide?
- Have we been guilty of genocide?
- What does it take to survive?
- How do the survivors cope?
- How can the world best bring justice to those who perpetrated genocide?
- Will these questions persist for centuries more?

Eleven thousand Jewish children from Paris

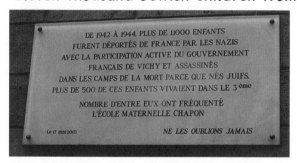

How could something like this happen? Just imagine! Eleven thousand children shipped off to concentration camps and slaughtered.

One day I was strolling through Paris enjoying a sunny afternoon when I passed L'École Maternelle Chapon and saw the plaque pictured above. "Let us never forget them," the plaque reads, placed there in May of 2003.

While we would like to think that such madness is a thing of the past and that genocide is practiced by others, current events offer plenty of examples of mass slaughter and nearly every country has been complicit at some time in some form of mass slaughter, including the slaughter of innocents.

Fear of darkness

One reason some schools may shy away from such issues and questions is the controversial and often troubling nature of the material. A month spent wrestling with such questions may result in little peace, almost no certainty and precious little resolution. Young ones can become troubled and distressed by these kinds of inquiries. There are few happy endings, in the Hollywood sense. There is a clear and present danger of disillusionment, angst and depression.

These same risks may partially explain how news programs can focus more attention on the antics of Paris Hilton than on the events in

Darfur.   Even the coverage of Iraq declined dramatically as time went on.   Certain news sells poorly and wins no audience.

The three TV broadcast networks' nightly newscasts devoted more than 4,100 minutes to Iraq in 2003 and 3,000 in 2004, before leveling off at about 2,000 a year, according to Andrew Tyndall, who monitors the broadcasts and posts detailed breakdowns at tyndallreport.com. But by the last months of 2007, he said, the broadcasts were spending just half as much time on Iraq as they had earlier in the year.

But some puzzles and enigmas are less dark and hardly troubling. They may not prove any more easily resolved, but these pleasant mysteries win little attention except in whimsical and fleeting dalliances as in an English class that considers a poem for a few minutes.   Poets have been fond of mysteries, both dark and amusing ones, for centuries, but a few moments of whimsey hardly prepare young ones for a lifetime of mysteries.

In Just—
spring when the world is mud—
luscious the little
lame balloonman

whistles far and wee

E.E. Cummings

What's it all about?

The recent obsession with reading and math in the USA turned attention away from serious problems with comprehension, understanding and thinking skills that have plagued us for decades when measured by the NAEP tests — the National Assessment of Educational Progress.

Unless we are teaching the young to wrestle with mystery and manage complexity, they are unlikely to meet the challenges of this century with skill and confidence.   A public that has been raised on the thin gruel of learning that was focused on simple test taking is unlikely to handle the tough ideas and concepts required for thoughtful citizenship during troubling times.   Raised on simple answers for simple tests, there is every danger that they will seek the same simplicity from their leaders - simple answers for complex problems.

Chapter 16 - What Digital Age?

Technology vendors and cheerleaders would have you believe this is a digital age, but some members of the society would have it otherwise, preferring a balance between things digital and the sensuous aspects of the natural world. Digital landscapes and menus are not always comparable. Shall we walk through a virtual rainforest or a real one? Hardly a choice unless you are addicted to the couch, the tube and your headphones.

Sadly, this is an age of substitutes, fakery and the ersatz, as "reality" TV shows and news magazines offer simulated experiences that too often disregard real news and the real world. To accept the digital label uncritically is a form of surrender to cultural trends that should inspire dissent and apprehension. Virtual reality?

While some aspects of the digital panoply are splendid, others are simply hollow. Is digital destiny?

Not surprisingly, the attempt by technology merchants to name this age "digital" goes pretty much unchallenged by the popular media, yet here and there a voice is raised to warn of the culture's drift away from the rich and the soulful to the thin and the barren.

Irish poet John O'Donohue, who passed away in January of 2008, devoted a section of his book *Beauty: The Invisible Embrace* to an analysis of how things digital may dilute experience and take us away from the beautiful.

O'Donohue explains, " . . . the digital virus has truncated time and space."

> The self has become anxious for what the next moment might bring. This greed for destination obliterates the journey. The digital desire for the single instant schools the mind in false priority. Page 27

He laments the impact of things digital on the quality of life:

> When you accumulate experiences at such a tempo, every-

thing becomes thin. Consequently, you become ever more absent from your life and this fosters emptiness that haunts the heart. Page 27

Such dire warnings stand in stark contrast to the claims of various purveyors of the Digital Age. According to them, life is grander, more fun and vastly more entertaining when it flashes by with the velocity of a TV ad. The cause of original thinking is poorly served by this trend.

Deep time and deep thinking

There are times when digital can enhance the beauty and meaning of life, as when good music and noise-cancelling headphones might filter out distracting sounds and voices while a writer struggles with important questions on a laptop. But a balanced view of things digital should include an honest look at the nerve jangling, interruptive phenomena that O'Donohue warns against.
As Sven Birkerts states . . .

> Resonance — there is no wisdom without it. Resonance is a natural phenomenon, the shadow of import alongside the body of fact, and it cannot flourish except in deep time.
> *The Gutenberg Elegies* — Page 75

Used wisely, digital resources can help us to find deep time, but they might also prove diversionary. The cell phone rings at many wrong moments. The email message arrives with a beep. An alert warns us of a stock loss or a governor's scandal. A string quartet lulls us into reverie.
It is a matter of choice and judgment. Each individual must learn how and when to go unplugged. Digital offers richness and value when chosen thoughtfully. We can hardly line up a live time, unplugged symphony orchestra when we hunger for the inspiration of Bach, Copeland or Mozart. Without recordings, how could we possibly hope to have such masterpieces at our beck and call? Similarly, how many of us can hop on a plane to enjoy Monet's Water Lilies in Paris? Without the rich offerings of virtual museums, all of lives would be lessened, no doubt.

Flying across the Pacific at 36,000 feet reading *The Spanish Bow*, a novel about a famous cellist, I realized I possessed and could hear on my laptop the Bach Cello Suites that figure significantly in this story

written by Andromeda Romano-Lax. Digital combined powerfully with print to enhance my appreciation of the music threading through the book.

Going unplugged: Running with or without headphones?

When race officials in the USA started to ban the use of headphones during races, the reaction of runners was quite mixed, as some runners never use headphones and argue that the incessant noise is disturbing as well as dangerous. Others find the music during a three hour run to be an important source of inspiration for thinking as well as running. The impact of music on mood and thought can be deep. Without doubt, certain music might prove better for running than thinking and the converse might also be true.

But how about running with email? When considering the complete array of digital experiences and resources, which are most likely to enhance deep thought and which most likely to disturb it?

Many of the telephone and technology companies push a digital lifestyle suggesting that it is best to be connected "any time and anywhere." Some of these ads promote the advantages of bringing one's work everywhere. Owners of Blackberrys sometimes complain that they no longer have a life - that intrusions are 24/7 and persistent.

One ad shows a woman executive enjoying the porch of a summer home in bright sunlight, sitting in dark glasses with her laptop and several other devices. The text reads "Spacious corner office redefined. Business is no longer confined by four walls. Today folks need to access and exchange information - anytime, anywhere." There is a zeal to intrude, invade and disturb - the selling of a work life that is ubiquitous and all-pervasive.

Life is a beach?

When things digital invade the beach, as we see in this cartoon, we should pause to consider the implications. Do we want a Digital Age? Children unable to sit at table without a Gameboy to keep them occupied?

There is something to be said for grains of sand between one's toes.

"Yeah, I got away from the office, finally and the beach is so peaceful. Can you hear the surf?"

Cartoon from http://phoneybusiness.com

Chapter 17 - A True Original

With important groups such as ISTE, AASL and the Partnership for 21st Century Skills all calling upon schools to stress originality, imagination and creative production, the meaning of those terms, along with related concepts such as synthesis, innovation and invention, becomes central to defining the purpose of education.

This chapter examines the meanings of the key concepts while considering how they might shape learning in schools during the next decade.

A true original?

When is a product, a work of art or an idea a true original and when is it a mere copy, a counterfeit or an imitation? When does copying cross the line into plagiarism?

"It is better to fail in originality than to succeed in imitation."
                                                    Herman Melville

How much originality can we expect from our students and how can we best nurture its development?

What does originality have to do with ICT?

The biggest failing of the past twenty years of ICT (Information and Communication Technologies) in schools has been the putting of carts before the horses, stressing the tools themselves rather than the information, the communicating and the building of new ideas and insight. This is a failure repeated across the culture and the society as fashionable and trendy uses often supersede thoughtful uses.

The focus of ICT should be on ways of knowing, learning and understanding, making smart use of information technologies to support exploration and contribute to both comprehension and invention.

ICT is about information, understanding and literacy — using thinking skills to make sense of the vast stockpiles of information now available electronically.

ICT is about smart thinking, not smart boarding.

Degrees of difference:
Shifting from the ridiculous to the sublime

1. Copying — When researching a challenge, the student simply cuts and pastes the ideas, statements and positions of others. Whether done with attribution or not (plagiarism), the student is simply gathering, not inventing or thinking.

2. Modifying — Rearranging, combining, condensing, summarizing and blending the ideas, statements and positions of others is akin to weaving or creating a tapestry. There may be a slight degree of originality in the rearrangement, but the student has taken gathering just one step further to amalgamation. The elements being synthesized did not originate with the student.

3. Improving — The student adds new insights and possibilities to those ideas gathered from others —- extending, elaborating, enhancing and elevating understanding to new levels, introducing elements that are quite novel to the mix in general circulation.

4. Starting Fresh — The student invests a great deal of thought to defining and mapping out the challenge at hand before collecting the insights of others. The student poses hypotheses and possibilities prior to collection. The research process is meant to develop understanding and construct ideas instead of collecting the ideas of others.

Who was a better poet?

If we ask our high school or university students to compare two poets, we should expect that they do the original thinking involved in such a comparison rather than scooping up the comments of some literary critic or critics.

Who was a better poet? Elizabeth Browning or her husband Robert?

Unfortunately, many students might be tempted to rush to the Net to find out what the sages have written on this question. They might soon find themselves reading the following passage from *The*

110

Cambridge History of English and American Literature:

> Her work, in fact, was as chaotic and confused as it was luxurious and improvident. Her Seraphim is overstrained and misty; her Drama of Exile is an uninteresting allegory; nearly all her shorter poems are too long, for she did not know how to omit, or when to stop. Few, if any, poets have sinned more grievously or frequently against the laws of metre and rime.

In this approach to the assignment, the student contents himself or herself with collecting and reporting the opinions of the sages. There is no original thought involved.

Original thought might begin with a cluster diagram sketching out the traits of a good poet.

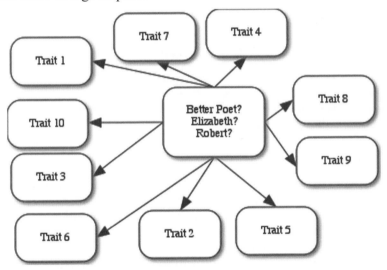

The student must wrestle with some difficult concepts. Does popularity matter? Income? Being widely published?

Creating new possibilities for poetry?

Sticking with the tried and true?

What makes a poet great?

Honoring conventions? Turning them upside down?

Being innovative? Being original? Being soothing? Speaking clearly but powerfully? Using rhyme? Writing poetry that withstands the test of time? Being colorful and lyrical? Being inspirational?

Ideally, each student will select different traits and each will create a cluster diagram that is unique - one that matches that student's preferences and values.

This approach to the assignment then charges the student with the responsibility of collecting data to build a case.

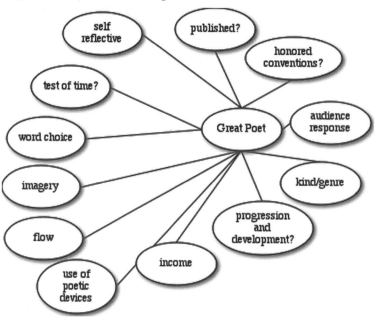

In comparing the poetic devices employed by Elizabeth and Robert (such as alliteration, assonance, imagery and metaphor — see http://www.dealapoem.com/poetic-devices.html), we should expect students to lay the Brownings' poems side by side.

The student might consider several of Elizabeth's *Sonnets from the Portuguese* such as the fourth, drawn from the online collection at http://www.bartleby.com or other sonnets from http://www.cswnet.com/~erin/ebbpoem.htm.

> IF thou must love me, let it be for naught
> Except for love's sake only. Do not say,

Robert's poems can also be found at Bartleby and at http://www.cswnet.com/~erin/rbpoem.htm.

> YOU'LL love me yet!—and I can tarry
> Your love's protracted growing:

The student should be developing a personal assessment of the power of each poet to capture and express complex ideas and emotions. The assessment should be grounded in terms of literary analysis and substantiated with specific lines and phrases as examples. It would be best if this assessment took place before reading the opinions of the sages.

This is not easy work. Gathering the opinions of others is much easier and involves much less analysis. To compare these two poets, the student must consider evidence besides their works.

- Which poet published most and was most read during their lifetime?
- Which poet is now published most and most read?
- Is the quality of a poet or a poet's career to be judged by their popularity? their acceptance by the general population?
- Is it reasonable to judge the poetry of those writing more than a century ago?
- How do standards of poetry change over time?

How original thinking readies students for the NAEP Reading Assessment

The kind of thinking suggested above matches that required to score at the Advanced proficiency level on the NAEP Reading Assessment as defined below for eighth graders:

Eighth-grade students performing at the Advanced level should be able to describe the more abstract themes and ideas of the overall text. When reading text appropriate to eighth grade, they should be able to analyze both meaning and form and support their analyses explicitly with examples from the text, and they should be able to extend text information by relating it to their experiences and to world events. At this level, student responses should be thorough, thoughtful, and extensive. Source http://www.nagb.org/publications/frameworks/r_framework_05/ch3.html

The Proficient level also requires the reader to build a case and understand literary devices and elements:

Eighth-grade students performing at the Proficient level

should be able to show an overall understanding of the text, including inferential as well as literal information. When reading text appropriate to eighth grade, they should be able to extend the ideas in the text by making clear inferences from it, by drawing conclusions, and by making connections to their own experiences—including other reading experiences. Proficient eighth graders should be able to identify some of the devices authors use in composing text. Source http://www.nagb.org/publications/frameworks/r_framework_05/ch3.html

How young?

The very young can be challenged to make up their own minds, form their own ideas and create fresh and original possibilities.

Original thinking should be thought of as a habit of mind, and primary schools should see their work as engaging students at the fourth level of the Degrees of Difference outlined above — Starting Fresh. In fact, the work properly begins in pre-school, as good teachers engage the very young in tasks that invite imagination, play, improvisation and wonder. Note article "In the Early Years."

http://questioning.org/jun08/early.html

This charge is well understood in New Zealand where the early childhood curriculum is clear about the importance of such themes. Note the article "The Early Childhood Workshop."

http://questioning.org/jun08/workshop.html

Chapter 18 – Play, Experimentation & Improvisation

"The play's the thing."

Shakespeare had something different in mind when he wrote those words for Hamlet to speak, but for this chapter, the words underline the importance of play in the early years of a child's life as a foundation for imaginative thought and production.

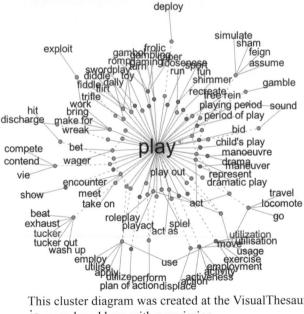

This cluster diagram was created at the VisualThesaurus.Com and is reproduced here with permission.

The disappearance of play

The current drift of culture in many countries actually works to undermine the vitality of play, especially free play According to a recent study, "Children's Pastimes and Play in Sixteen Nations: Is Free Play Declining?" by Dorothy G. Singer, Jerome L. Singer, Heidi D'Agostino, and Raeka DeLong in the Winter 2009 issue of the *American Journal of Play*, "A decline in opportunities for free play and experiential learning is eroding childhood around the world."

Through two years of interviewing, we hear the voices of mothers from around the world calling out in Spanish, Portuguese, Turkish, Thai, and other languages. . . . deeply concerned that their youngsters are somehow missing out on the joys of childhood and experiential learning opportunities of free play and natural exploration.

Even though a number of organizations such as the American Association of School Librarians, ISTE (the International Society for Technology in Education) and the Partnership for 21st Century Skills are calling for schools to stress original thought and imaginative production, schools in the United States and many other countries have been caught up in handling the demands of high-stakes testing in ways that might block such goals.

The connection between play and invention

Creative production is directly tied to a spirit of playfulness without which it is very difficult to generate exciting new possibilities. Imagination thrives when the spirit of play is cultivated and encouraged.

Many of the organizational development strategies designed to promote more inventive thought focus on promoting the attitudes required to test out promising new variations and possibilities.

Roger von Oech, suggests in *A Whack on the Side of the Head*, that we suffer from mental locks that inhibit our creative production:

Play is Frivolous	The Right Answer	I'm Not Creative
That's Not Logical	That's Not My Area	To Err Is Wrong
Follow the Rules	Don't Be Foolish	
Be Practical	Avoid Ambiguity	

116

To unlock the potential of groups, von Oech introduces a number of games and activites to counter the influence of the mental locks listed above.

The Association for Childhood Education International has published a position paper with the title, "Play: Essential for All Children," that makes a strong case for play:

> Theorists, regardless of their orientation, concur that play occupies a central role in children's lives. They also suggest that the absence of play is an obstacle to the development of healthy and creative individuals. Psychoanalysts believe that play is necessary for mastering emotional traumas or disturbances; psychosocialists believe it is necessary for ego mastery and learning to live with everyday experiences; constructivists believe it is necessary for cognitive growth; maturationists believe it is necessary for competence build-ing and for socializing functions in all cultures of the world; and neuroscientists believe it is necessary for emotional and physical health, motivation, and love of learning.
> Source: http://www.acei.org/playpaper.htm

The Alliance for Childhood has launched a major campaign to counter a number of disturbing trends:

> Among these are the loss of creative play and hands-on ac-tivities in children's lives, and the excessive amounts of time spent in front of screens instead of in face-to-face engagement with other children, caring adults, and the natural world. We also work against the commercialization of childhood, the misuse of high-stakes testing, and increasing levels of child-hood obesity.

The group has published a report, "Crisis in the Kindergarten: Why Children Need to Play in School."

> New research shows that many kindergartens spend 2 to 3 hours per day instructing and testing children in literacy and math—with only 30 minutes per day or less for play. In some kindergartens there is no playtime at all. The same didactic, test-driven approach is entering preschools. But these meth-ods, which are not well grounded in research, are not yielding long-term gains. http://drupal6.allianceforchildhood.org/publications

The American Academy of Pediatrics has released a report, "The Importance of Play in Promoting Healthy Child Development and Maintaining Strong Parent-Child Bonds" that states firmly that play is essential:

> Play is essential to development because it contributes to the cognitive, physical, social, and emotional well-being of children and youth. Play also offers an ideal opportunity for parents to engage fully with their children. Despite the benefits derived from play for both children and parents, time for free play has been markedly reduced for some children. This report addresses a variety of factors that have reduced play, including a hurried lifestyle, changes in family structure, and increased attention to academics and enrichment activities at the expense of recess or free child-centered play.

Penn State has provided a wonderful article, "What Will the Children Play With?" outlining eight areas of play and how they can be set up to stimulate and encourage play.

1. Creative Art
2. Large Muscle Activities
3. Small Muscle Activities
4. Language Activities
5. Dramatic Pretend Play
6. Construction/Building/Blocks
7. Music
8. Science

http://betterkidcare.psu.edu/AngelUnits/OneHour/
EquipmentMaterials/EquipMatsLesson.html

Starting with parents

Given the cultural drift away from play, it would be wise for schools to start with parents to make sure they understand the value of play — especially unstructured or "free" play. Much of the shift in childhood experience for the young has been engineered by parents who are anxious to provide their children with wonderful opportunities in the arts, in sports and in a host of areas where they see development aided by formal and structured learning opportunities. It is a rare parent these days who understands the importance of free play, but they can be won over when they are shown the connection between free play and innovation.

A number of articles might prove useful in launching such a parent education program. One published by *Scholastic*, "The Joys of Doing Nothing," by Margery D. Rosen, makes a convincing case for unstructured play.

"This generation of parents has swallowed whole, and in some cases, is choking on, the belief that the sooner you expose a child to learning, the more he or she will learn," says Rosenfeld. "And if they don't get it during those critical early childhood years, well, forget Harvard."

http://www2.scholastic.com/browse/article.jsp?id=1450

Redesigning homework

When schools have piled on many hours of homework each night, they have been complicit, so a review of homework policies is a good place to start. While some kinds of homework are beneficial and likely to promote imaginative thought, others are busy work unlikely to contribute to learning. Schools should limit the time devoted to homework and focus on those activities with the highest payoff in terms of student learning and growth.

In addition to shifting homework to stress valuable activities, the school should limit the number of hours each night so the child ends up with some free time after doing homework and various after school classes. Here again the parents become important, because free play would not include TV viewing or computer games. Given a list of twenty choices, parents should be helped to understand which undermine and which strengthen the development of imagination.

Time to breathe

Within the school day itself, children should be given ample opportunities to play, provided with significant breaks throughout the day at recess and lunch so that their spirits can revive and can take flight. Proponents of extra instructional time often try to seize these minutes away, claiming that scores will rise as a result of extra learning time, but so far there is no convincing evidence that this works, and even if it did increase reading and math scores, the price paid would be the loss of imagination combined with stress and anxiety. The biological, physiological and psychological arguments for recess and breaks are compelling. Human learning thrives when batteries are given a chance to recharge. Students return to class primed for the next dose of learning.

The National Association of Early Childhood Specialists in State Departments of Education has issued a position paper, "Recess and the Importance of Play," that outlines the benefits of recesss along with the research substantiating those benefits.

The position paper goes into considerable detail with regard to how recess influences the social, emotional and physical development of children. http://naecs.crc.uiuc.edu/position/recessplay.html

Making Room for Experimentation

Playing with blocks is an old standby in kindergarten programs that can pay off mightily in terms of development, especially if the young children are inspired to experiment and see how high a tower or how far a bridge they can build. Many of the areas of play outlined earlier by Penn State provide examples of places where the young can practice experimentation. Sadly, as children move into test-driven stages of elementary school and beyond, the focus often shifts to carefully structured learning activities tied to narrowly defined learning objectives and test items. The drive for accountability has often led to a slimming down of the curriculum and a thinning of the learning opportunities.

Unfortunately, because there is a relationship between reading comprehension and experimentation, the focus of reading upon lower-order tasks and patterns threatens the development of advanced comprehension skills. When students encounter the most difficult comprehension questions, they cannot rely upon memorized patterns and strategies. These questions are often open ended and require inference. Students must read between the lines, play with clues, test hypotheses and act a bit like a detective.

In recent years we have seen that some states have achieved what looked like miracles in terms of student achievement so long as that achievement was measured by state tests, but those same states did very poorly when their students took the NAEP tests.

The eighth grade reading questions on the NAEP test require an experimental bent — the capacity to wonder and wander a bit, playing with the evidence until an answer emerges.

Below are a few examples of such questions:

1. Do you think the lesson in this story is true today? Why or why not?

2. Explain what makes this story a fable?
3. What was the major character's opinion of _____? Use evidence from the story in your response.
4. What causes the main character to do _____? Use evidence from the story in your response.
5. How do you think the character's actions might be different today? Support your response with evidence from the story.

The rest can be found at http://www.nagb.org/publications/ frameworks/r_framework_05/ch2.html

Experimentation involves the testing of theories and possibilities. It requires an attitude as well as a skill set. When reviewing the research by P. David Pearson on reading comprehension, this capacity to try out different strategies in different combinations might be called "flexing" as outlined in F below:

A. Questioning — A proficient reader employs a toolkit of questions in order to solve puzzles, unlock mysteries and fashion meaning when it is elusive.

B. Picturing — A proficient reader relies upon her or his mind's eye to increase understanding and organize complex ideas.

C. Inferring — A proficient reader can read between the lines, put clues together and figure out reasonable interpretations.

D. Recalling Prior Knowledge — A proficient reader awakens relevant memories of information that might cast light on the topic or issue at hand.

E. Synthesizing — A proficient reader is skilled at combining fragments and ideas in new ways so as to squeeze import and novelty from them.

F. Flexing — A proficient reader knows how to slide back and forth across an array of strategies until one succeeds, picking and choosing purposefully.

Source: "Power Reading and the School Library"
http://www.fno.org/sum05/powerread.html

Much of the material above was drawn from the work of P. David Pearson and his research on reading comprehension in the 1980s identified strategies of proficient readers — work cited by Stephanie Harvey in *Nonfiction Matters: Reading, Writing and Research in Grades 3-8*. More recently, Pearson has combined with others to determine which approaches to change have the most impact "on schools with populations of students at risk of failure by virtue of poverty." *Teaching Reading: Effective Schools, Accomplished Teachers*. Edited by Barbara M. Taylor and P. David Pearson. Lawrence Erlbaum Associates, Mahwah, New Jersey, 2002. Chapter Sixteen — "Research-Supported Characteristics of Teachers and Schools that Promote Reading Achievement" — is particularly relevant to the thrust of this chapter.

As we saw in Chapter Four,, for the young to make their own meanings and figure things out for themselves, they need to be able to shift back and forth across the four operations of wondering, pondering, wandering and considering. Each has its own role and its own time, but often they will operate concurrently.

When reading passages, understanding depends upon this dynamic

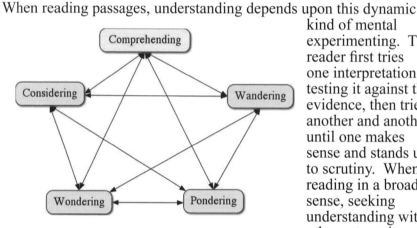

kind of mental experimenting. The reader first tries one interpretation, testing it against the evidence, then tries another and another until one makes sense and stands up to scrutiny. When reading in a broader sense, seeking understanding within other categories of information as outlined in Chapters Twenty-One and Twenty-Two of this book, this experimental approach remains a core practice — the prime method to grasp elusive meanings.

Daily experiments

For students to become well versed and comfortable with experimentation, it should be a part of every day's schooling and learning.

One can achieve such a goal in several ways. One approach is the use of problem-based learning strategies so students are grappling with questions like the ones below:

- What do you suppose went wrong in the past?
- What are the key forces and variables operating here?
- Which ones offer the most opportunity and leverage?
- What do you suppose would happen if we combined the three most promising of the strategies we brain-stormed and researched?
- What kind of data must we gather to assess the success of our plan once we launch it?
- What do suppose might go wrong?
- Is there any way to prevent that from happening?

But experiments can also be identfied within the regular curriculum with a lot less fanfare.

Students can learn to experiment or play with words as they write poems, testing out different variations on Web sites that support magnetic poetry.

1. *Magnetic Poetry*
http://www.magneticpoetry.com/magnet/
2. *Shocked Poetry*
http://www.shockedpoetry.com/

When students apply the *Six Traits of Effective Writing* to their drafts, they are also experimenting with voice, word choice and organization. See "From Start to Finish" at http://www.fno.org/sept98/infolit5.html to see how this might operate.

Whether studying concepts in science or in social studies, opportunities abound to engage students in wondering and testing what might happen if certain things are changed around. In a novel the teacher might ask students to predict the ending if a certain event turns out badly.

Opportunities abound to make experimentation a daily challenge so long as teachers recognize the importance of this challenge and seize those chances to make it come to life.

Improvisation and all that jazz

For the purposes of this chapter, the concept of improvisation is linked to a number of other words that all capture some aspect of thinking on your feet, reacting to a surprise challenge without much preparation and performing well despite the pressure.

Photo of Jazz quartet by Jamie McKenzie

The publishing world offers dozens of books suggesting ways to think on your feet, but the concept has not received much attention in the educational world.

Here are just a few examples of the many books:

- *Thinking on Your Feet: How to Communicate Under Pressure* by Marian K. Woodall, Professional Business Communications; 2nd edition (June 1996)
- *Improvise This!: How to Think on Your Feet so You Don't Fall on Your Face* by Mark Bergren, Molly Cox and Jim Detmar. Hyperion, March, 2002.

In many respects, we expect students to learn how to ad lib when caught in a high pressure and surprising situation, but we would also hope that their response, while unrehearsed, would sitll be rooted in sound thinking, past experience and a firm grasp of the fundamentals.

In the world of jazz, young performers must master a repertory of chord progressions and harmonies so that they can count on them as structures around which and through which they might weave more magical variations.

It is much the same with the improvisation we should require of students in school. We ground them in the fundamentals and then ask them to stand up and deliver. We are not satisfied with canned presentations. We hand them a card that gives them a role and a position to defend. We give them a few minutes to prepare and then we send them to the podium. They are good at thinking on their feet.

Chapter 19 - Professional Development
for Inventive Learning

Original thinking and synthesis have not received much attention during the formal training of most teachers I have met during my travels in the past few years. I have asked perhaps a hundred audiences to recall whether they ever received explicit instruction during university or graduate school in how to perform synthesis — as much as thirty hours or more. In most audiences no hands go up. Occasionally there will be one or two hands per hundred sitting there.

In nearly every country, province or state I visit, the governments have listed original thinking, creativity and synthesis as major program goals, yet there has been little organized in those places to close the gap between the wonderful intentions of the planners and the capacity of teachers to deliver those goods.

Closing that gap will require more than a one-day workshop or seminar. Most teachers will require between five and ten of such days to acquire the skills to nurture the originality and inventiveness of their students.

During the past year while writing this book, I offered a one-day seminar devoted to the concepts of originality, synthesis, invention, imagination and novelty in more than 20 cities. In a five hour seminar, we could just scratch the surface. It was an appetizer, really, not the full-course dinner teachers really deserve and would need to turn the rhetoric and good intentions into effective daily practice.

If the educational planners are true to their intentions, they must follow up the declarations and proclamations with resources to fund extended and effective professional development. They must also consider how assessment efforts might be aligned with such program goals so that teachers will see their efforts recognized and rewarded. So long as there is a dearth of resources and a poor fit between tests and goals, we are unlikely to see a major shift to translate these goals into practice.

Failing dramatic leadership and funding from the top, it would still be possible and worthwhile for individual schools or groups of teachers to take on this challenge for the good of their students and themselves. This chapter suggests fruitful strategies for such efforts.

Professional development
as organizational development

The most effective learning strategies require a change in the ways teachers spend their time and the ways they work together. Frequently we notice how informal support systems, partnerships, teams and collaborative structures may be the most important elements in a broad-based change effort.

Gardening provides a useful metaphor for this process. We will see more growth if we cultivate the soil and fertilize before planting. An exclusive focus on skills is a bit like spreading seeds across a concrete playground.

While some maintain that reluctance to use different techniques is rooted in a lack of skill and confidence, teachers need to be recruited. They must be convinced of the value of the new activities and then given ample time to work on teams to invent effective lessons.

In many schools, teachers are isolated from each other and preoccupied with what Michael Fullan calls "the daily press" of getting through their schedule, focused in many places on creating good (high) test scores.

Quite a few of these teachers are likely to cling to routines they have enjoyed in the past until they are equipped and encouraged to find, invent and test new routines that are suitable and reliable replacements.

This creative exploration, invention and testing will require a change in schools that breaks down isolation, facilitates the work of teams and provides ample time for program development. Such efforts also require a commitment to the project from school leadership that is unequivocal.

Professional development
as adult learning

What do we mean by "adult learning" and how does it differ from the training models that have dominated many staff development offerings over recent decades?

The clearest way to contrast adult learning (often called "andragogy") with pedagogy (instructor directed learning) is to note that adult learning usually involves the learner in activities that match that person's interests, needs, style and developmental readiness.

Fundamental beliefs:
- The learner may make choices from a rich and varied menu of learning experiences and possibilities.
- Learners must take responsibility for planning, acting and growing.
- Learners will more readily embrace that which is chosen freely.

If we shift school cultures to support adult learning, professional development is experienced as a personal journey of growth and discovery that engages the learner on a daily and perhaps hourly basis. In the best cases, andragogy includes an emphasis upon self-direction, transformation and experience. One learns by doing and exploring — by trying, by failing, by changing and adapting strategies and by overcoming obstacles after many trials.

Unlike the training models, adult learning is primarily concerned with creating the conditions, as well as the inclination and the competencies, to transfer new tools and skills into daily practice. While training usually occurs outside of context and frequently ignores issues of transfer, adult learning is all about melding practice with context. Adult learning should encourage teachers to identify and then remove obstacles.

Peer coaching

Peer coaching represents an especially promising way to implement adult learning strategies, as teachers work with trusted partners to test out different strategies while supporting each other through the challenges.

Joyce and Showers were able to profile the elements in a successful teacher change process through field studies that demonstrated that skill acquisition thrives when peer coaching and support systems are in place to encourage risk taking. Their research demonstrates that participants with peer coaching partners are more likely to persevere with the introduction of innovative strategies than isolated individuals. Even though past practice possesses momentum and inertia, partnering has proven effective in pushing new practice through the often troubling and frustrating incubation stages. Showers, Beverly & Joyce, Bruce "The Evolution of Peer Coaching" in *Educational Leadership*, March, 1996.

The study group

Given the enormity of the challenge, teams of teachers, perhaps 5-6 per group, might form around the purpose of developing synthesis skills by meeting once a month around shared reading and classroom efforts. Akin to the peer coaching suggested by Joyce and Showers, this approach engages a group in reading, discussing and testing over an entire year or more.

Carlene Murphy (1998) and others have developed and tested study group models that fruitfully engage every teacher in such activities. Her work stands out as she seeks a method to engage all teachers in a school rather than a select few. http://www.murphyswfsg.org

A book like this one might serve as the focus for a study group's efforts, or another good choice might be Michalko's *Thinkertoys: A Handbook of Creative-Thinking Techniques.* Michalko's book is 394 pages long. It explains several dozen techniques in some depth. He devotes pages 71—108 of *Thinkertoys* to SCAMPER, a synthesis strategy designed in the 1950s to help business people come up with new products and ideas.

Each of these strategies might be worthy of a month's trial, as members of the study group first discuss and then test the strategy with students in activities appropriate for their teaching assignments. After trying the synthesis strategy, the group meets again to talk over what worked and what did not work.

It is quite clear from just a partial listing of the strategies in Michalko's book that the challenge of developing a comfortable grasp of these techniques would take several years.

Linear Thinkertoys	**Intuitive Thinkertoys**
False Faces	Chilling Out
Slice and Dice	Blue Roses
Cherry Split	The Three Bs
Think Bubbles	Rattlesnakes and Roses
SCAMPER	Stone Soup
Tug-of-War	Color Bath
Idea Box	Dreamscape
Idea Grid	Da Vinci's Technique
The Toothache Tree	
Phoenix	

The personal plan

Considering the personal nature of this learning journey, another promising approach to professional development places the focus and the responsibility on individual teachers, with each mapping out prior to the beginning of a school year the steps necessary to learn and then incorporate these techniques into the daily classroom experience.

This kind of plan might be a personal sketch reserved for the teacher's eyes alone, or it might be a more formal document expected as part of a supervisory system that calls for teachers to set goals shared formally with the principal or some other supervisor. The choice of goals might be entirely up to the teacher, or in some cases, some of the goals will be assigned as school-wide goals. There are quite a few variations possible, but the important thrust of this approach is the personal planning and commitment expected and encouraged. Every effort is made to maximize the sense of ownership and investment.

Sustainable professional development

When it comes to teachers learning and valuing the effective use of the teaching techniques outlined in this book, schools will discover that the kinds of training programs offered in the past may not represent the most generative method of reaching a full range of teachers and their students. The key term is "generative" — meaning that behaviors and daily practice will be changed for the better as a consequence of the professional development experience and that those changes will endure — be sustainable — over decades.

There have been too many bandwagons over the years that passed through schools without ever really touching down or winning take-up. These short term fads actually distract teachers from the real business of school without doing much good for anyone.

Teaching students to produce original ideas is a worthwhile goal, but there is little chance it will endure unless a school digs in and devotes 4—5 years to the learning journey.

Not long ago, I was presenting a workshop on questioning. A gentleman of my own (advanced) age interrupted me rather abruptly, "We've done thinking three times in my career. What are you selling that is any different?"

His question was on target. How can we best introduce programs so they take root, sprout and blossom? A commitment by a school to the kinds of adult learning described here might be one good answer.

Chapter 20 - Building a Case

Instead of relying upon the conclusions and reasoning of others, we expect students to develop their own positions based on evidence and sound logic. This process might be called building a case.

We are accustomed to thinking of detectives and lawyers building cases against criminals or rogue corporations, but doctors must build a case for a treatment plan and inventors must be able to persuade someone to translate their idea into a product.

Building a case is central to persuasion, sales and advocacy.

In this chapter, we look at six ways to engage students in such thinking.

A debate

Resolved: The U.S.A., Australia and other Coalition nations should stick with the Iraq War until the job is done.

Over the decades, many students have learned to build and to deconstruct a case as part of their preparation for debate. They must gather all the arguments on both sides of the resolution. On the one hand, they must construct a persuasive presentation outlining the case for their position, but they must also anticipate all the ideas likely to emerge from the opposition team. Once they build these lists, they must unmask the weaknesses in both cases. They turn this knowledge into a competitive advantage, strengthening their own case by bolstering the evidence or the logic while figuring out how to make the other side look foolish.

A proposal for action

The state government should take the following action to protect the coast:

Strategy 1? Strategy 2? Strategy 3?

When students explore problems from society and make suggestions for their resolution, they pass through several stages that move from definition of the problem, through research of possibilities, and then on to the development of an action plan listing steps that might be taken. This selection of particular measures should be bolstered by a rationale — a clear statement explaining why these steps amount to the best response. Once again, students are presenting evidence that argues the wisdom of a particular course of action. At the same time they assert the case for the options selected, they should dispose of other possibilities by listing their weaknesses or inadequacies.

Many adult teams are afflicted with wishful thinking when they develop a plan, failing to consider evidence that might conflict with their preconceptions or hunches. Students must learn to consider all dimensions and possibilities rather than starting with an idea and only noticing that which seems to match their original position. An action plan should be grounded in reality, and the team developing the proposal should be concerned with its efficacy. Sometimes leaders confuse ideology and faith with wisdom, charging along a disastrous path without sufficiently testing those beliefs against reality.

Problem-based learning strategies (PBL) are intended to give students a taste of decision-making applied to challenges from the past or present, albeit, a taste that is simulated.

Online resources explaining PBL abound. Some of these can be found at http://fnopress.com/pbl2/toc.html.

An indictment

Predatory Bakery is guilty of unethical business behaviors and practices.

A.I.G. executives who persisted in taking extravagant trips after the first bail-out were guilty of extraordinary arrogance.

George Custer was guilty of war crimes.

Sitting Bull was guilty of war crimes.

Current events and history afford many opportunities for students to ponder the guilt, responsibility or innocence of various actors on the world stage. But it is not enough to make judgments. The students must also garner and weigh sufficient evidence to sustain a measured and well-considered conviction, avoiding the various intellectual traps that afflict many who attempt to pass judgment. While it is tempting sometimes to act on bias and preconceptions and rush to a conclusion, the process should be more deliberate, careful and restrained. Even though the case may seem simple on the surface, there are often extenuating circumstances that deserve attention. The search for balance, evenhandedness and justice may prove frustratingly slow and demanding in comparison with a rush to conviction, but lynchings, literal or figurative, violate our sense of decency.

As with all of these examples of building a case, students will benefit from study of trials in the past that might illustrate the difficulty of finding justice, whether these be drawn from history (i.e., the Scopes Trial) or from literature (i.e., *To Kill a Mockingbird*).

A judgment

George W. Bush was a better president than his father.

Making well-founded judgments is a skill that will serve our students well as they move through adolescence into early adulthood. Whether they are comparing ship captains, cities, fast food restaurants or poets, the process challenges them at the top level of Bloom's *Cognitive Taxonomy*. They must first establish criteria and then find evidence to substantiate their judgment.

- What are the traits of a good captain?
- What are the traits of a good poet?
- What are the traits of a good president?
- What are the traits of a good fast food restaurant?

Once students have agreed upon criteria, they develop "telling questions" for each category like the three listed below for the French fries (or chips as the Australians would call them.

- What is the fat content?
- What is the taste?
- What is the texture?

To determine healthy levels of fat and compare the fries of one restaurant with those of the others, students must do some research. These issues can be defined numerically if the restaurants will provide nutritional data, but some will refuse, in which case students may have to conduct experiments to develop their own data.

Taste and texture may be a bit more difficult to define, but students can develop rating systems that are then applied in blind taste tests or touch tests like those conducted by Pepsi and Coke to compare their products.

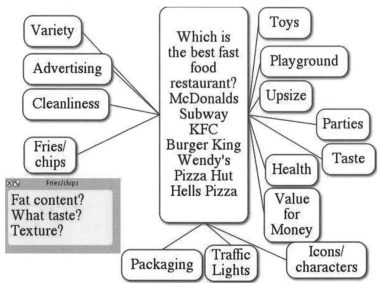

The main point behind these exercises is the importance of gathering data to substantiate a finding.

Judgments normally apply to things past, while choices, as outlined in the next section, apply more to the future. The lines become blurred, since good choices are based on sound judgment.

Students may normally rely upon their gut instincts when selecting fast food restaurants, but an evidence-based approach has much more merit when selecting a leader or a course of action.

A choice

Of all the careers that beckon to me, which are most likely to offer me the most satisfaction and the best life?

The thought process required to make sound career choices is similar to that just outlined for reaching a judgment. Once again, the student must pay careful attention to definitions and criteria.

- Just what is a good life?
- What gives me the most satisfaction?
- Will my preferences change as I age?
- What are the main elements in a job or career?

- Which elements would I care most about?
- Can I list a dozen criteria that would serve me well in thinking about and contrasting careers?

Unlike the fast food example, there are more unknowns when making choices about the future. It may be more difficult to nail down specifics and collect hard or definitive data. In some respects, this decision-making involves some of the complexity and mystery outlined in Chapter Four. One cannot know with much assurance how one's preferences and desires will change with age and experience.

The listing of criteria and the collection of evidence should help to guide the choice, but many people must taste the new life before they can actually determine how well suited they are to the experience they have selected. Quite a few finish law school and start practicing law before they figure out it is not satisfying. Others may love the law but find the demands of a top firm unreasonable. They may change course and start their own less-lucrative private practice that allows them to exercise more control over hours and life style. Some may begin as prosecutors but switch to the other side after trying a mix of criminals and innocents. Others may start as public defenders but turn to corporate law after years of frustration and low pay.

Making smart choices is especially complicated when it comes to romance, as humans are remarkably vulnerable to wishful thinking and various forms of infatuation that may blind their judgment. But it is also difficult to select new employees despite the best attempts to set criteria and gather data. It is simply not an entirely scientific process. Surveys, interviews and background checks cannot deliver certainty.

The attempt to impose rational order upon emotional and subjective domains must be seen as a somewhat fallible process, but one that is still worth pursuing.

An invention

We have drafted legislation to make sure that all workers are adequately paid and fairly treated.

Many inventions never pass from the idea stage to real production, whether they be a physical product or an improved way of doing things. Often this failure is tied to the poor marketing and persuasion skills of the inventor. Those who can imagine new possibilities do not always have the capacity to build a case for the new thing.

There are many worthwhile invention projects that engage students in dreaming up new machines or concepts. An important aspect of such projects should be the marketing of the invention.

- Why do we need this better mousetrap?
- What will it do for us?
- Is it worth the price?
- Has it been adequately tested?
- What could possibly go wrong?
- How might it be improved?

If possible, those who have a new law in mind or a new product ready for production should have gathered data to prove the importance and effectiveness of their proposal. It makes little impression on venture capitalists when someone steps forward with something new that is not validated by some kind of market research as well as a business plan outlining how the product will eventually make a big profit. Before they back a new proposition, they will require a convincing presentation outlining the benefits bolstered by evidence and well developed strategies. Likewise, legislators who seek majority approval of a new bill must provide convincing arguments as well as evidence to win those votes away from competing bills.

Chapter 21 - Reading Between Digital Lines

There is often a subtext when it comes to digital media and information, even though that subtext may be quite well hidden and difficult to identify. Not all that matters is glimmering on the surface.

Inference is required now more than ever. Students must know how to read between the lines, put clues together and fathom meanings that are not self-evident. They must look past the words.

The very term "reading" must be applied more broadly than it was in the past to include the capacity to comprehend any category of information or medium, as in "read a face," "read a building," or "read a situation." Students must now understand how to "read a page" — understanding implicit meanings, context and the intentions of the information producers. They become astute with regard to page design and subliminal manipulation.

As an increasing share of the texts young people read are now delivered electronically, the challenge of comprehension has shifted in some ways from the days when most words appeared in books, magazines, newspapers or notes passed in class.

As outlined in Chapter Five, students must now learn to cope with the "poverty of abundance" — overwhelming amounts of information — as well as issues of credibility and authority. Reading skills of prior decades remain important but they are not sufficient.

Reading a page

Lots of money is now spent on page design — how the information on a page is displayed and arrayed. Many of these aspects of communication will be subtle and subliminal unless students have learned to consider the techniques employed on each page. Compared to pages in a book from years past, the layout of the page and its visual components may have great impact on one's understanding of the information.

Google offers a dramatic example, as its basic page is remarkable in its minimalist design while its pages reporting search results are crammed with information carefully arranged so that "sponsored links" appear around the so-called regular search results in subtle ways, as colors and placement are used to make them pretty much unnoticeable. 89 millions search results for laptops! What a great example of the poverty of abundance — a mountain of pages that sets up *Google* to make money by steering visitors to sponsored links. The information tidal wave actually increases dependency.

Students as prospectors

For students to read between digital lines, they must become prospectors capable of finding the gold, the oil or the water below the surface. *Roget's Thesaurus* draws a connection between the act of prospecting and a treasure hunt. But it may prove to be a treasure hunt through a garbage pail or landfill!

The information prospector must . . .

Scour, clean out, turn over, rake over
Pick over, turn out, turn inside out
Rake through, rifle through, go through
Search through, look into every nook and cranny
Look or search high and low
Search high heaven
Sift through, winnow, explore every inch

S*Roget's Thesaurus of English words and phrases.*

For this kind of digging to be successful, students must be skilled at asking probing questions.

Probing questions

In previous articles published in FNO and subsequent books, this type of question was outlined in detail. It is one of a dozen question types that students must learn to employ in combinations. They are described in "The Questioning Toolkit" at http://fno.org/nov97/toolkit.html.

Probing questions take us below the surface to the "heart of the matter." They operate somewhat like the archeologist's tools — the brushes that clear away the surface dust and the knives which cut through the accumulated grime and debris to reveal the outlines and ridges of some treasure.

Another appropriate metaphor might be exploratory surgery. The good doctor spends little time on the surface, knowing full well that the vital organs reside at a deeper level.

And when we first encounter an information "site," we rarely find the treasures lying out in the open within easy reach. We may need to "feel for the vein" much as the lab technician tests before drawing blood. This "feeling" is part logic, part prior knowledge, part intuition and part trial-and-error.

Logic — We check to see if there is any structure to the way the information is organized and displayed, if there are any sign posts or clues pointing to where the best information resides. We assume the author had some plan or design to guide placement of information and we try to identify its outlines.

Prior Knowledge — We apply what we have seen and known in the past to guide our search. We consider information about the topic and prior experience with information sites. This prior knowledge helps us to avoid dead ends and blind alleys. It helps us to make wise choices when browsing through lists of "hits." Prior knowledge also makes it easier to interpret new findings, to place them into a context and distinguish between "fool's gold" and the real thing.

Intuition — We explore our hunches, follow our instincts, look for patterns and connections, and make those leaps our minds can manage. Especially when we are hoping to create new knowledge and carve out new insights, this non-rational, non-logical form of information harvesting is critically important.

Trial-and-Error —ß Sometimes, nothing works better than plain old "mucking about." Push here. Tug there. Try this out! We find a site with so much information and so little structure that we have little choice but to plunge in and see what we can find.

Debunking and deconstructing

In these times of spin, deception and false claims, debunking and deconstructing have become survival skills. Things are often quite different from what they appear to be.

When we accept claims and reports at face value there is a substantial risk that we will be hoodwinked. Variants on hoodwink include the following list from the thesaurus:

> deceive, trick, dupe, outwit, fool, delude, inveigle,
> cheat, take in, hoax, mislead, lead on, defraud,
> double-cross, swindle, gull, scam; con, bamboozle

Good teachers equip students with the questioning skills to challenge the veracity of information, whether it appear on a library site, a "green" site, a politician's site or pages advertising a product.

Does the "free range" chicken actually venture outside or is the label a misleading technicality? The door was left open and the chickens elected to remain inside. They may still carry the label "free range" on their package.

Can we trust the pharmaceuticals to reveal the real risks?

Why did the incidence of breast cancer decline dramatically when the use of hormone supplements dropped off?

"Get real!" is a rallying cry for those who wish to cut through the hype, the exaggeration and the pseudo-science that is often used to promote an idea, a product or a plan of action.

Given the prevalence of distortion, skepticism is warranted when anyone is pitching their wares, whether they be manufactured goods, news, government policies or candidates.

Sadly, any expectation of verity is naively old-world these days as simulacra and the plastic often substitute for the real. Virtual reality shoves the other reality into the background, as the hip version is often more attractive and entertaining than the original.

Program suggestions to build these skills can be found at The Center for Media Literacy (CML) - a nonprofit educational organization that provides leadership, public education, professional development and educational resources nationally. http://www.medialit.org

The Center outlines a program called "Five Key Questions That Can Change the World: Deconstructing Media" at http://www.medialit.org/reading_room/article661.html

Unlike other media literacy activity books, which typically are organized by genre (news, advertising, etc.) or topic (violence, gender, etc.), the inquiry-based lessons in *Five Key Questions That Can Change the World: Deconstruction* help students build an internal checklist of questions to ask about any message in any media — television, movies, the Internet, radio, advertising, newspapers and magazines, even maps and money.

Reading?

As mentioned in the previous chapter, we need a broadened conception of reading to capture the many different types of reading that occur when considering information in different formats across different media. Students must now be able to "read a face" as well as a page, must be able to read a photograph or a chart or a situation. Reading as understanding applies to many aspects of life.

What is the mood communicated by the statue on this grave in the famous Cemetiere du Pere Lachaise in Paris?

What details in the way she is sitting communicate the mood?

What details in her face communicate the mood?

Why do you suppose someone wanted this statue on top of their grave? Would you?

A dozen literacies

What is a literacy?

For this book we will define *literacy* as the capacity to analyze, interpret and understand information within a particular category of information or within a particular medium. Each category may require specific tools, concepts and vocabulary to unlock the full meaning of the information provided.

While currently there is an international trend to narrow schools' focus upon just two literacies — textual and numerical — that trend is fundamentally undemocratic and disturbing in its implications. An informed person capable of voting intelligently on the difficult issues of the day must be capable of understanding much more than words and numbers. This chapter suggests a dozen categories of information, each of which involves a literacy. Good schools will accept the challenge of equipping students to understand information in any and all of these categories.

Natural Literacy

Do we still know how to "read the forest" and its many signs of life? Do we recognize the various signs of life? Back when many humans relied upon such reading to find meat for their families, such skills were basic skills. Fewer people hunt for survival now and many never make it out of the city, even if they do drive a rugged "all terrain vehicle," but sport hunters still view such reading as a basic.

It takes special skills to read a swamp or a beach or a desert area. These skills also differ from region to region as the flora and fauna shift.

Most of us have heard of swimmers caught in rip tides because they did not know how to read the signs, or of visitors enjoying tidal flats suddenly swept up in an incoming tide much larger than anything they knew back home.

As with many of the literacies mentioned in this chapter, there can be overlap or linkage between natural literacy and environmental or scientific literacy. Some suggest that citizens might vote for greener policies if they were more in tune with nature and had not lost their appreciation for its wonders as well as its vulnerabilities.

For a wonderful further explanation and exploration of natural literacy, read "Natural Literacy" in WaterNotes at http://www.seanursery.com/water/60

Artistic literacy

Anyone can look at a painting, a photograph or a movie.

Anyone can comment on a piece of art.

Anyone can listen to a symphony or sit through an opera.

But looking, commenting, listening or sitting do not automatically translate into understanding.

One can learn to read a photograph — understand its elements and interpret its meanings. Students should acquire a set of skills as well as a vocabulary that strengthens their analysis. Aesthetics empower the viewer to look more carefully and thoughtfully at the work.

There are many resources available on the Web to support the development of this kind of artistic literacy, whether it be the capacity to look at photographs or listen to a symphony. Kentucky Educational TV, for example, offers a lesson plan and video, "Abstraction—Critique of Art" at http://www.ket.org/painting/abs-critique.htm.

> Students will observe and make personal decisions
> about abstract artworks using a four-step critique pro-
> cess:
>
> 1. Description
> 2. Analysis
> 3. Interpretation
> 4. Judgment

ArtsEdge at the Kennedy Center offers an excellent page "Teaching Students to Critique" by Joyce Payne at http://artsedge.kennedy-center.org/content/3338/ that expands on the four-step process mentioned above.

Each artistic medium presents its own special interpretive challenges as well as special vocabulary and aesthetics.

As with other literacies, there is a good deal of overlap or connection between artistic literacy and others such as visual literacy and media literacy, as an understanding of photography and music is obviously part of deconstructing and analyzing the marketing and propaganda so prevalent in today's media.

Media literacy

Media literacy involves the capacity and the inclination to cut past the distortions and manipulation often typical of today's news, communications and entertainment media in order to build an understanding of the world that is at least partially grounded in reality.

With the conglomeration of these industries, the lines between news and entertainment have been blurred.

1. Disney
2. News Corporation
3. Sony Corporation

4. Time Warner
5. Viacom
6. Vivendi Universal

To see the operations owned by these huge corporations go to http://www.internet-web-directory.com/Business_and_Economy/ Entertainment_and_Arts/Media_Conglomerates/index.html

The same company that reports the news on TV often makes films, produces textbooks, generates student tests, operates Web sites, prints newspapers and offers reality TV shows.

Media literacy was fully outlined in Chapters Eight and Nine.

Ethical literacy

The collapse of financial institutions for making fortunes on dubious loans and associated scandals such as Wall Street financier Bernard Madoff's alleged $50 billion fraud highlight the need for increased awareness regarding the ethical dimensions of decisions. A society that celebrates the leadership styles of Donald Trump and Martha Stewart by giving them TV programs (*The Apprentice*) shows little appreciation for the ethical ironies involved in rewarding an ex-con with more attention and celebrating Donald's offensive ways of handling workers, treating "You're fired!" as a clever joke.

Many schools have embraced problem-based learning as a way to involve students in wrestling with real world challenges — an excellent way to introduce them to the moral and ethical aspects of modern life. A similar approach is called Authentic Learning — developed by Fred M. Newmann and Gary G. Wehlage. "Five Standards of Authentic Instruction."

Students are involved researching important issues, problems and challenges. As they move from "What can be done?" to "What should be done?" the ethical considerations come into play.

Several groups have posed ethics as an important aspect of learning. One organization, the Institute for Global Ethics (IGE), is "dedicated to promoting ethical action in a global context. Our challenge is to explore the global common ground of values, elevate awareness of ethics, provide practical tools for making ethical decisions, and encourage moral actions based on those decisions."

Their Web site at http://www.globalethics.org/ offers a list of dilemmas drawn from a wide range of domains:

- Business Dilemmas
- Education Dilemmas
- Children and Family Dilemmas
- Medical Dilemmas
- Philanthropy Dilemmas
- Personal Dilemmas
- Military Dilemmas

The Institute for Global Ethics offers both services and resources aimed at schools or other organization, ". . . to help you put ethics into action."

Unlike the other literacies, ethical literacy usually involves aspects of situations that are embedded, implicit or lying below the surface, so a major step is the identification of the moral dimensions of issues and decisions.

Visual literacy

As mentioned earlier, there is some overlap between visual literacy and artistic literacy, since visual literacy involves the analysis and interpretation of images, but not all images are artistic in nature. They may communicate important data in chart form, for example. They may be photographs of a current event — a natural disaster or invasion.

"What's the story here?" is a good lead question for images of this kind. The students must learn to identify the important content, some of which may require inference.

There is also some overlap between visual and media literacy, as students consider the adequacy and veracity of news images. As Israel sent troops into Gaza in January of 2009, which images were distributed by each side in the conflict and which ones were ultimately viewed in countries like the U.S.A., Australia, China and Iran?

Inflation has hit the image business big time as we must edit the old saying from a thousand words to a million or a billion words. But sometimes a missing image is worth even more. "Out of sight is out of mind." If we cannot see "collateral damage," maybe it didn't happen? If we have no image of torture by water (waterboarding), maybe it isn't so bad?

Numerical literacy

In many schools too little attention is devoted to the use of numbers and data to help students understand their world. Unfortunately, some folks are highly skilled at distorting truth with numbers and with charts. Numbers often seem to have face validity. There is a saying that "Numbers don't lie," but it should be revised to read, "Numbers don't lie, but people lie with numbers."

We have seen the wreckage that results when people distort truth with numbers as they did in the mortgage and financial industries. Measures of financial health and worth were fudged. Some firms cooked the books.

There are some excellent Web sites designed to teach students how to consider these issues, such as *Numeracy in the News*, a site in Australia that encourages students to critique the soundness of reasoning and evidence contained in newspaper articles summarizing recent studies. http://www.mercurynie.com.au/mathguys/mercindx.htm

There are questions for students like the following at the end of each article:

1. Do you think that the information provided in this article justifies the claim in the headline and first paragraph? Why or why not?
2. Is it possible that there could be any other factor (a lurking or hidden variable) which could be causing both drinking beer and violent crime? Discuss some possibilities.
3. Explain the difference between two variables being correlated and one causing the other to vary. In this article correlation is used to infer causation. Is it reasonable? Why or why not?

This kind of critical thinking about data is a critical skill for this century as many groups try to influence voting and consumer behavior with studies that are often quite flawed.

Text literacy

Even though many governments are stressing text and numerical literacy, they often set the bar too low and fail to emphasize challenging levels of comprehension. In all too many cases, the measures of student competency seem designed to make the government look good instead of asking whether students can unlock the mysteries of difficult texts.

The sample items on the NAEP clearly require that students figure

out meanings for themselves rather than merely find answers in the text. The emphasis is on inference and reading between the lines.
http://www.nagb.org/publications/frameworks/r_framework_05/ch2.html

This kind of reading is much more demanding than items calling for students to locate answers within the text. Many of the NAEP items are open-ended — requiring students to write answers rather than select them from a multiple choice listing.

Social/cultural literacy

Many nations like the U.S., Australia, New Zealand and Canada are finding that immigration has bestowed upon them a wide range of traditions, ideas, and customs from other lands. Many consider this mix a blessing — an enriching source of much beauty and inspiration — while some bemoan the newly arrived and their ways as strangers and interlopers. Ironically, trace the roots of these folks back far enough and it often turns out their own ancestors were once viewed as encroachers, as a recent exhibit by Native Americans pointed out. Aptly named "Trespassers," the exhibit at the Whatcom County Museum of Art and History explored the impact upon native life and customs when Europeans first arrived in Indian lands.
http://www.whatcommuseum.org/pages/exhibitions/exhibitions.htm
Years ago, newcomers were pretty much expected to blend into the dominant culture. The U.S.A. proudly called itself a "melting pot," but in recent years some have suggested that "mixed salad" might be a better metaphor. Rather than eradicating the customs of newcomers from China, Pakistan and Mexico, why not celebrate them and ask how these traditions and talents might strengthen the nation? Rather than wave good-bye to their languages of birth, why not encourage their retention so they can help the nation compete in the increasingly global market?
All students must learn to read the behaviors and customs of these various groups and must blend a degree of tolerance and respect into their behavior when in contact with others, whether it be as students in a school with diverse groups, as workers or as community members. Not all customs may be acceptable, as might be the case with dog fighting, cock fighting or the mistreatment of women and others, but many other customs might be worth emulating. Some groups pay tremendous respect to elders, for example, a tradition that has been eroded in some so-called Western cultures. The very notion of beauty

is quite different in some cultures and those countries that embrace these variants may find the arts thriving.

Emotional literacy

Emotional literacy and intelligence have received much popular attention in recent years as several authors have built a strong case for knowing how to read the feelings of the people in one's life. Sensitivity and empathy are seen as essential elements in building relations and working effectively with other people, whether they be colleagues, friends, family members or opponents of some kind. Because much human decision-making is swayed by emotional reactions, those who fail to develop this literacy are likely to stumble and blunder in their interactions with others.

Understanding the feelings of others involves a complex array of reading skills — the capacity to read facial expressions, body language and many other cues with which people communicate, often unconsciously, their attitudes. Many of these skills can be learned. And then again, folks can learn to reduce such signals when it serves their purpose, as is the case with "poker faced."

For further reading on emotional literacy go to the *EQ Alliance* at http://www.eq.org/ — a listing of resources.

For further reading on emotional intelligence, consider *Emotional Intelligence: Why It Can Matter More Than IQ* by Daniel Goleman.

Organizational literacy

How well can the consultant read the leadership team she has been hired to lead through a strategic planning process?

How well can the new superintendent read the leadership team she has been hired to lead?

How well can the new teacher read the expectations of his new school and his new boss? Is it ok to have great bulletin boards? Is it ok to stay in the building into the evening hours? Is it ok to speak up at staff meetings?

Even kindergarten children must learn to read the organization.

- Who is the boss here?
- What are the rules?
- What are the real rules?
- How can we avoid punishment?
- Who is the class bully?

- Who is the class victim?
- Who is the class clown?
- How can we get rewarded?

Many organizations use climate surveys of various kinds to judge the health of their group while identifying aspects in need of attention.

Literacy involves understanding, and in the case of organizations, the goal is to use that understanding to shift or reinforce values, actions, beliefs, norms and sanctions so that the goals of the group are advanced rather than bogged down or subverted.

Environmental literacy

As organizational literacy will focus upon the health of the group, environmental literacy keeps an eye on the health of the planet, asking what we must all do to promote that well being and work to promote sustainability. There is some obvious overlap with scientific literacy - the one that comes next in this chapter — because understanding the environment requires a grasp of many important scientific principles and concepts.

Scientific literacy

For a long time, environmental literacy was subsumed under scientific literacy, just as poetic or musical literacy might be grouped under artistic literacy, but the urgency of threats to the ecology have heightened concern and awareness so that it deserves a separate, special place in anyone's list. The umbrella term "scientific literacy" captures the need for all citizens to grasp the important issues and concepts arising out of a number of spheres such as chemistry, biology, geology, physics, and astronomy.

One recently formed group, The Foundation for Scientific Literacy, asks the following questions:

- How do we reconcile the technological marvels in our lives?
- Do we understand them?
- Do we blindly accept them?
- Do we trust the scientists that bring these advances to us?
- Or do we question them?
- Do we understand well enough to ask the right questions?
  http://www.scientificliteracy.org/

The American Association for the Advancement of Science has published extensive resources to support the development of scientific literacy.

Atlas of Science Literacy, Volumes 1 and 2
Mapping K–12 science learning

The Atlas of Science Literacy is a two-volume collection of conceptual strand maps—and commentary on those maps—that show how students' understanding of the ideas and skills that lead to literacy in science, mathematics, and technology might develop from kindergarten through 12th grade.
    http://www.project2061.org/publications/atlas/default.htm

Many reports lament the sad state of science literacy in the United States. In her editorial in Science, 14 August 1998: Vol. 281. no. 5379, p. 917, Jane Maienschein suggests that we consider both science literacy and scientific literacy:

> The first emphasizes practical results and stresses short-term instrumental good, notably training immediately productive members of society with specific facts and skills. We call this science literacy, with its focus on gaining units of scientific or technical knowledge.
>
> Second is scientific literacy, which emphasizes scientific ways of knowing and the process of thinking critically and creatively about the natural world. Advocates of the second assume that it is good to have critical thinkers, that scientific literacy is an intrinsic good — on moral and other principled grounds. Being scientifically literate helps people to live "good" lives (in the philosophers' sense of reflective and fulfilling, and not in the distasteful sense of eating good-for-you bran flakes).

http://www.sciencemag.org/cgi/content/summary/281/5379/917

Nexus?

1 [noun] the means of connection between things linked in series
Synonym: link
2 [noun] a connected series or group

WordWeb dictionary

For two decades, schools have brought new technologies into classrooms without fully appreciating the importance of literacies and comprehension as the driving forces that could make their investments pay off in the form of enhanced student performance. In many cases, schools fell into the trap of doing technology for the sake of technology.

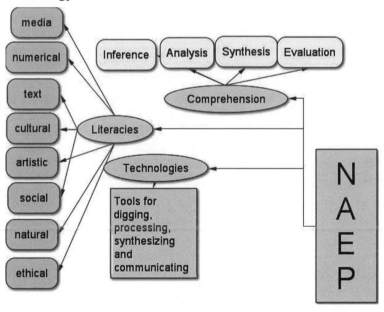

This chapter explores the promising connections between technologies, literacies and comprehension as ways of equipping students to manage challenging tests such as NAEP (The National Assessment of Educational Progress) as well as the even more important tests of life.

NAEP - The toughest tests in the land

NAEP (The National Assessment of Educational Progress) has been around for decades, but it was not until the advent of NCLB that states were required to submit a sample of students to its scrutiny. It has always been a challenging collection of tests with a minority of students achieving proficiency levels, but it acts as a powerful standard against which to measure the validity of state proficiency claims. Some of the states claiming miraculous progress are shown by NAEP to be frauds.

Sadly, NCLB offers many loopholes so that some states can make their students look good on state tests without them actually doing well. The NAEP tests hold all states accountable in a way that closes the loopholes.

When we look below at the huge gaps in some states between claims of student proficiency and actual proficiency as measured by NAEP, it is quite clear which states have the biggest gaps.

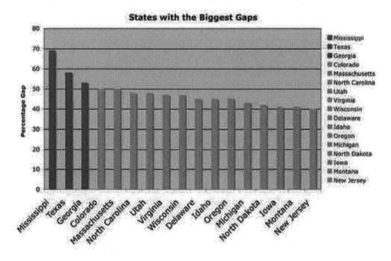

States with the Biggest Gaps

Is this real Reform?
Source: "The Performance Gap" at
http://www.nochildleft.com/2005/jan05gap.html

152

Proficiency requires thinkers
Not automatons!

What is a thinker?

And how could students spend lots of time in school doing thinking without ever becoming thinkers?

A thinker is someone capable of generating fresh ideas, building insight, making inventions and figuring out how to confront problems. A thinker does not rely only on conventional wisdom and is rarely content to cut and paste the thinking of others.

Research rituals

A student can spend weeks studying an important person from history or literature without actually doing any important thinking, yet it might seem to the casual observer that the student has been thinking long and hard.

Remedy: Questions of Import

- In what ways was the life remarkable?
- In what ways was the life despicable?
- In what ways was the life admirable?
- What human qualities were most influential in shaping the way this person lived and influenced his or her times?
- Which quality or trait proved most troubling and difficult?
- Which quality or trait was most beneficial?

You will find an expanded version of the list above at *The Biography Maker* - http://fno.org/bio/biomaker.htm

Avoiding another brick in the wall

We don't need no education.
We don't need no thought control.

Thinkers do more than smush!
Collecting and smushing lots of information is a type of synthesis.
But we should expect more of our students.

Chapter 24 - One Liners, Bloggery and Tomfoolery

Technologies work in curious ways.

The surface magic sometimes conceals sinister currents.

The way we speak to each other, the way we write to each other and the way we exchange messages and ideas — all of these are shifting. In many ways these changes represent progress, but there is also a dark side that is rarely noticed or mentioned. It is easy to surrender to the trendy, the new and the popular without questioning value.

As schools consider ways to nurture the original thinking of students, they must also keep an eye on ways that new technologies might subvert such goals lest they inadvertently participate in the decline of language and thought.

Lessening

One ominous trend is the potential thinning or lessening of warmth, content and consideration brought on by text messaging. Abbreviation alternates with voluminous chit-chat as the order of the day.

The ease and speed of connectedness is both a blessing and a curse.

We can dash off a line in text to a distant friend and know it will circle the globe in nanoseconds. We can call when stuck in traffic. We can post pictures of a new grandchild on the Web. We can read strong opinions on blogs that would never make it to the mainstream media. We can subscribe to email newsletters that illuminate, and we can enjoy podcasts by remarkable people as we jog along through a forest.

These are blessings, for sure, but sometimes we sacrifice quality for convenience. Dashing, rushing and chatting can dilute the depth and thoughtfulness of the communication.

We may find one liners replacing carefully crafted paragraphs or even pages of text. Niceties and sensibilities often suffer as compression trumps style and subtlety.

Messaging vs. dialogue

During the past ten years or so, messaging has begun to eclipse dialogue as a way to communicate. Ironically, cell phones make it easy to chat endlessly with good friends, while at the same time, messaging has allowed screening and distancing from others who are less favored.

In many cities you will see drivers and strollers with cell phones glued to ears.

Stuck in traffic and alone in a car? The cell phone allows connectedness when isolation threatens.

In some cities walkers plunge ahead while text messaging at high speed. Not so long ago this level of contact was unimaginable.

The business traveller relies on a mobile phone to report movements:

"We just landed."
"We're approaching the gate."
"I'm heading down to baggage now."
"I'm on the rental van."
"I'm getting into the rental car."
"I am heading North on the Interstate."
"What's up?"

Blah blah blah

Some of this exchange is trivial chatter. Some of it is vital.

"I'm stuck on a plane that will be delayed so that I miss my connection. What can you do to protect me?"

"We're stuck in traffic and the baby is coming any moment now. Can you send a helicopter to get us?"

At the same time these kinds of contacts have multiplied dramatically, a large share now represent the leaving of messages rather than dialogue. Some people never answer their home phones, especially at night, wary of telemarketers or fund raisers. Caller ID gives us the power to downgrade a caller to mere messaging status. Although we may answer the phone for special people, many others are effectively shunned.

How is dialogue different? What do we lose with messaging?

The advantage of good dialogue is the chemistry of interaction — the back and forth, real time consideration of ideas that allows for construction, development and insight. In contrast, messaging usually amounts to the leaving or dropping off of ideas or beliefs. Messaging may support less productive weaving and synergy.

Phone tag

Phone tag?

While we joke about this experience, it is a symptom of social and cultural shifts brought on by the new technologies. If we hunger for a real conversation, why is it so hard to reach people when we need them? Given the impressive array of new tools, how could it be so frustrating?

It is easy to shrug off this phenomenon by blaming it on busy lives and busy people, but there is more to it than that.

Some people restrict their cell phones for personal use and a narrow group of friends with whom they wish to be perpetually connected. They may answer their cell during a meal in a restaurant if a special friend calls and then sit chatting while their dinner companion is ignored. This same person may be hard to reach through their work or other land lines - the ones they share with those outside the inner circle.

The blunt one liner

Text messaging gives birth to very short messages.
Niceties are eliminated
Nuance is throttled.
Elegance is sacrificed.

"Cya!"

In contrast, poetry is a short form that provides richness and intensity, while a text message is rarely poetic or lyrical. It is most often dashed off with little thought or attention to style. A poem is a heavily reduced, flavorful sauce. Text messages are most often thin and watery, leaving much to the imagination.

Acronyms and abbreviations abound:

ADIP = Another Day In Paradise
CRTLA = Can't Remember the Three-Letter Acronym
DILLIGAD = Do I Look Like I Give A Damn
GMTFT = Great Minds Think For Themselves
ILICISCOMK = I Laughed, I Cried, I Spat/Spilt Coffee/Crumbs/
Coke On My Keyboard
Source: List of Acronyms & Text Messaging Shorthand at
NetLingo - http://www.netlingo.com/emailsh.cfm

"R U OK?"

The "Age of Glib"

For more than a decade, I have warned of trends toward super-ficiality and glib production as new technologies have made it easy to scoop, smush and flash. Note article, "The Mind Candy Kafe" at http://www.fno.org/jun98/kafe.html

Does this Mind Candy Kafe presage the arrival of an Age of Glib?

If students always surf along the surface, will they ever learn to think for themselves?

Will they settle for the sound bites, the mind bytes, the eye candy and the mind candy offered up by the media like sticks of chewing gum?

Or will they develop a healthy skepticism about the informa-tion (and noise) streaming past them?

In many cases, the mountains of information combine to obscure meaning and delay the search for understanding. Without a dramatic commitment to the teaching (and practice) of information literacy skills in school, our students will be ill prepared to find their way.

Bloggery

Blogs are a mixed blessing.
At their best, blogs deliver startling information that might other-wise never reach us through traditional media. Some bloggers write with style and careful thought. Their blogs are essays posted for gen-

eral consumption in ways that circumvent traditional circulation and publication channels. As such, they do the society a service, widening and (in some cases) deepening the public discourse.

At their worst, blogs are blather.

Not everyone has opinions or prose worth reading. *MySpace* makes it easy for everyone to post their own blogging, much of which is nothing more than *bloggery*.

> ***Bloggery*** - Blather and boring comments that some blogger enters in a blog without much thought, consideration or style. Blah blah blah posing as meaningful commentary.

> Example: "Her newspaper columns are much better than her blog which slips into bloggery of the worst kind."

As with many technology fads, the proliferation of blogs is impressive in the early stages, but once the first blush has passed, other issues arise.

- Is there anything worth reading?
- Where do we find the time to read so many pages of first hand, rambling comments?
- How much chit chat and blather can the human mind tolerate?
- What is the importance of style, content and thoughtfulness?
- What are the dangers of posting private thoughts?
- When do blogs cross over the line into slander and harassment?
- When is a blog a blog and when is it an article pretending to be a blog?

Most importantly, if no one reads a blog other than its author and one or two relatives and friends, does it really exist? How does it rank in contrast to graffiti that may be seen by hundreds?

References

Books and Articles

Baudrillard, Jean. *Selected Writings*, ed. Mark Poster. Stanford; Stanford University Press, 1988, pp.166-184.

Berliner, Paul F. *Thinking in Jazz*. University of Chicago Press, 1994.

Birkerts, Sven. *The Gutenberg Elegies*,

De Bono, Edward. *Six Thinking Hats*

Harvey , Stephanie. *Nonfiction Matters: Reading, Writing and Research in Grades 3-8*

Jacobs, Vicki A. "Adolescent Literacy: Putting the Crisis in Context" *The Harvard Educational Review*, Spring 2008.

Keen , Andrew. *The Cult of the Amateur: How today's Internet is killing our culture*

Michalko, Michael. *Thinker Toys: A Handbook of Creative-Thinking Techniques* (2nd Edition). Ten Speed Press, 2006

O'Donohue, John. *Beauty: The Invisible Embrace*

Payne, Joyce . "Teaching Students to Critique" at http://artsedge.kennedy-center.org/content/3338/

Postman, Neil and Weingartner, Charles. *Teaching as a Subversive Activity*

Prensky, Marc. "Digital Natives, Digital Immigrants." On the Horizon (NCB University Press, Vol. 9 No. 5, October 2001)

Rosen, Margery D.. "What Will the Children Play With?" http://betterkidcare.psu.edu/AngelUnits/OneHour/Equipment-Materials/EquipMatsLesson.html *Scholastic*, "The Joys of Doing Nothing," http://www2.scholastic.com/browse/article.jsp?id=1450

Senge, Peter. *The Fifth Discipline*.

Senge, Peter. *Schools That Learn.* New York: Doubleday, 2000.

Showers, Beverly & Joyce, Bruce. "The Evolution of Peer Coaching" *Educational Leadership*, March, 1996.

Singer, Dorothy G. et. al. "Children's Pastimes and Play in Sixteen Nations: Is Free-Play Declining?" *American Journal of Play,* Winter 2009.

Solnit, Rebecca. *A Field Guide to Getting Lost*

Taylor , Barbara M. and Pearson , P. David. *Teaching Reading: Effective Schools, Accomplished Teachers.* Edited by. Lawrence Erlbaum Associates, Mahwah, New Jersey, 2002.

References

Turkle , Sherry. *Life on the Screen*: *Identity in the Age of the Internet.*

von Oech , Roger. *A Whack on the Side of the Head*

Reports

AASL Standards for the 21st-Century Learner
http://www.ala.org/ala/mgrps/divs/aasl/guidelinesandstandards/learningstandards/standards.cfm

The Alliance for Childhood. *Fool's Gold: A Critical Look at Computers in Childhood.* October, 2000.
http://www.allianceforchildhood.net/projects/computers/computers_reports.htm

Alliance for Childhood, "Crisis in the Kindergarten: Why Children Need to Play in School."
http://drupal6.allianceforchildhood.org/publications

American Academy of Pediatrics "The Importance of Play in Promoting Healthy Child Development and Maintaining Strong Parent-Child Bonds"

The Association for Childhood Education International. "Play: Essential for all Children," http://www.acei.org/playpaper.htm

ISTE, (International Society for Technology in Education) *NETS for Students* 2007.
http://www.iste.org/AM/Template.cfm?Section=NETS

Kaiser Family Foundation. "Generation M - Media in the Lives of 8–18 Year-Olds." March 2005. http://www.kff.org.

Kentucky Educational TV. "Abstraction—Critique of Art" http://www.ket.org/painting/abs-critique.htm.

The National Association of Early Childhood Specialists in State Departments of Education "Recess and the Importance of Play" http://naecs.crc.uiuc.edu/position/recessplay.html

The Partnership for 21st Century Skills
http://www.21stcenturyskills.org

Index

Index